PRICE'S PROGRESS

PRICE'S PROGRESS

The Tortuous Journey
of a Roving Civil Engineer

James Price

VANTAGE PRESS
New York / Los Angeles

Let my book record for posterity my way of life and my gratitude. To all those I knew; they made my life worth living.

FIRST EDITION

Copyright © 1990 by James Price

Published by Vantage Press, Inc.
516 West 34th Street, New York, New York 10001

Manufactured in the United States of America
ISBN: 0-533-08620-5
Library of Congress Catalog Card No.: 89-90355

Contents

A Drama at Thirty-nine Holywell Road, Flint (1917)
To my mother, Gwendoline (Bithell) Price

The young girl—restless, Mother by her side
Sister—anxious, worried, but defied
The urge to sleep, to slip away
In dreamland to another day.
The heaving body—panting on the bed
Thro' the long night so dreary and so dead.
"When will it come?" they asked each other now.
What will it be? What shape and shade and how
Could it resemble them? The beauty of that girl
Heaving and panting; thinking's all awhirl.
Then when it seemed she could not bear it more
Looking so haunted, pale, wan, and sore,
They looked at her so gently and they said,
"When the mail train comes through 'twill wake the dead."
Soon, 'ere the time drew near to 3:00 A.M.
Her water broke and she became so tame.
Pushing was easier, labour near the end
Looking so meekly to those who did attend
And when the whistling, rumbling of "the mail"
Came through, out came its head. She heard the wail.
The girl lay back; knew she had given birth
And I became another soul on earth.

PRICE'S PROGRESS

1. The Light of Day

At 3:00 on a cold Tuesday morning, February 6, 1917, a girl of seventeen gave birth to a bouncing nine-pound boy child, while her young lance corporal husband was serving his country with the South Lancashire Regiment.

The address in the ancient borough of Flint in North Wales has long since been demolished. The soldier passed over in 1976 at the age of eighty-two, but the girl, Gwenny Bithell, survives as I write this in 1982. Happily, so does the baby, for it was I who first saw the light of day in a tiny two-up-two-down, end-of-terrace cottage at 39 Holywell Road near the gasworks.

Ma's eldest sister, Violet, informed James Jones, registrar, who duly registered the signally important event for your author, the birth of James Price, son of Horace and Gwendoline P.

Before joining up, my dad had been a plasterer journeyman, and during the war my ma helped eke out his few bob a week by keeping a small fruit store. Before her early marriage she had worked in Thomas's bakehouse. Ever since I can remember she had a way with cakes and icing. For his part, my dad was always a real artist with ornamental ceiling and cornice work. He also had a good practical head on his shoulders, a big one, on which he wore his black, wavy hair parted in the middle when he was young. Later his pate was just like those of all the Price males, thin to bald on the top and thick on the back and sides; they were real monastic heads.

Nain and Taid, Eliza and Tom Bithell, had six children: Vi, Gwen, Doris, Ste, Tommy, and Annie, who was only five years my senior.

Grandad and Grandma, Charles and Hannah Price, had nine offspring to my recollection: Thirza, Jim, Charlie, Harry, Nell, Ernie, Horace, Hilda, and Kathleen.

Uncle Jim was known to everyone in Flint as Jim the Ship because he kept the Ship Hotel. In my very early days it was there that Ma and I lived in a small upstairs room with a double and a single bed, a smoky fire, an oil lamp, and a washbasin and jug on a marble-topped stand.

We had our meals with the others who lived in the pub in the dining room downstairs.

Auntie Hilda and her husband, Uncle Ike (Jones), helped Uncle Jim. They, too, lived in the Ship. Also living there were little Nommus (enormous, only small, like a dwarf) and old Maggie; they did things like window cleaning, scrubbing floors, dollying the clothes, or putting sawdust in the pub's spitoons. Ike could be down in the cellar or ironing the billiard table or gluing on the cue tips, while humming "Beautiful Dreamers," "I'm Forever Blowing Bubbles," or some such popular song.

My first sortie into the literary world was to spell *Bass* and *Worthington*, words I could see painted in gold letters on glass wall advertisements. It was Uncle Ike who taught me to spell the words.

One of my earliest memories is the day Tommy came over to the Ship to see Ma and me upstairs in our room. Two or three ladies were also in the room, which was noisy with chatter and smoky from the kettle on the hob of the bedroom fire.

Tommy announced shyly to the female majority, "I'll take Jimmy over to his Nain's."

Ma gave me a big hug when Tommy asked me to kiss her before we left. I am sure she and I had not been apart before.

Out in the street we were walking past Raven Square when Tommy asked, "How would you like a little brother or sister?" I had not thought about it. I was only two and a half.

That day I stayed at Holywell Road with my Nain and the others. It was such an intimate home with its fire burning in the kitchen grate, the black-leaded oven with a high fender and fire irons, and a cupboard in the corner on the left. The kitchen also contained a sideboard; a tall chest of drawers; a sofa under the window looking onto the backyard, which led to a narrow garden; and a square table with four legs and three upright chairs to keep the sofa company around the table.

The stairs were semiconcealed and led from the kitchen to the two upstairs bedrooms.

In the front bedroom Nain slept alone, Taid usually being on his boat, the *Pilot,* which plied between Merseyside and Cardigan Bay carrying coal, stone, or iron ore.

The back bedroom had two double beds. Ste and Tommy slept in one and Doris and Annie in the other. At my young age I alternated to whichever bed had accommodation when I stayed with Ma's people.

On this first occasion, after two or three days, Tommy said he was taking me back to Ma.

On our short journey home we passed Cassidy's Sweet Shop on our left and the Raven Hotel and chip shop on the right. Then as we passed Pritchard's the barbers on our left, just before getting to the Ship, Tommy warned me, "You've got a little sister."

Inside our upstairs bedroom a baby was crying, while a lady in white washed her face and bare body in the washbowl.

The Ship Hotel, Flint, in 1986. The original Ship Hotel was owned by Uncle Jim Price, known as Jim the Ship. This was the author's home from 1917 to 1922.

I ran to Ma, who was in bed, and she hugged me tightly. Auntie Hilda was also in the room. "You've had your nose pushed out," she said to me.

I found myself looking into a hand mirror. "No, my nose is still there, Auntie," was my reply.

Only after one or two others had spoken about my nose did Ma explain it was a funny saying when a new baby came to live in a house, but it did not mean there was anything to worry about.

So August 22 being Iris's birthday, probably the earliest recollections I retain are of about this time in 1919.

Well, my dad came to see us only now and again. Ma told me he would be happy to see our new baby. Already we called her Iris.

Dad was home for a few days after Iris was born and much fuss was made of my new sister, but one day I managed to pull him into the backyard to look at old Grumpy, the pig, who had little piggies in the sty house with straw on the floor. We threw in some fine pieces of coal slack to go with the pig swill from a bucket Dad put into the trough, and Grumpy crunched it as if it were sweets.

Then Dad said, "I'm going back today, son. We'll see if Pritchard is busy."

Pritchard and Benny had a hairdressing shop with a window looking toward the pigsty in our backyard. Dad put a piece of wood across the big water butt below the tiny window, stepped onto the wood, and went through for a haircut.

I was still watching the pigs when I heard a splash and a shout. It was Dad up to his chest in the cold rainwater in the old black tub.

"That devil Nommus," he said, but he thought it was quite funny that the wood had been removed by the little man, keen to keep everything tidy. Dad just laughed and went to change. Then Dad had a meal and put on his khaki uniform with his flat-peaked cap and brass buttons, his putties, his black shiny boots, and his kit bag. He was late for the train so quickly ran through the front door of the Ship, and we all watched as he tossed his kit bag over the railway wall, took a leap up commando-style, and mounted onto the top and over to get onto the platform just in time to catch the train.

The men in the pub (there were no women customers) used to be tickled to see me toddle into the bar with Gypsy, my tiny Yorkshire terrier puppy, and often gave me a penny, sometimes even threepenny joeys, and these I secreted away in my "money box." This was a hole I found in the boards of the bare wood floor behind the bar counter. There must have been hundreds of coins discovered by the builders who virtually demolished the old pub and rebuilt the premises as the new Ship Hotel some years later.

On a Christmas Day I was sitting on the kerb outside the pub front door. The old town hall stood in front of me across the road, which surrounded the public building, with the main railway station behind on the left.

A bigger boy, about seven years old, was watching me with my new coloured ball in one arm, Gypsy in the other. He held my ball as I showed him my new knife from Father Christmas, all bright and shiny. He took the pocketknife and I had the ball again as he picked up a piece of wood from the gutter and started to whittle away.

"I'll just walk around the town hall," said the boy. When he came back he told me another boy had taken the knife from him. I was very sad at the loss, but glad I would never see the boy again.

Uncle Tommy Bithell was only eight years older than I, but when he and I were together he was very protective. When he heard of the loss of the knife he was most sympathetic. He said, "Perhaps I can find something else nicer than a knife."

I was in Nain's house when he came home one day on his bike from the county school, Holywell. He had with him a brand new magnet, a shiny red U-shape with shiny metal tips and a piece of metal across the tips. He had also brought some fine metal powder called iron filings and showed me how to make all kinds of patterns on white paper with the magnet underneath the paper.

"What did you eat for your dinner, lad?" I heard Nain asking him quietly in the back kitchen. "Two penn'orth of chips, Ma," was Tommy's answer. It was later I understood he had gone without his school "four-pennies" that day in order to buy his nephew a new surprise gift.

This act of Tom Bithell was a signal of his love and devotion to the service of others. It was appropriate for him to live all his life in the town and with the people of Flint. One year in the sixties those people made him mayor of Flint. To me, he was top of the town from his very early days.

One day little Nommus took me to the shops on an errand. There were a lot of people crowding the pavement. I was on the inside, and Nommus held my hand tightly.

Big Foulkes was walking towards us, with other people milling around us as they walked. Then suddenly Big Foulkes collided with little Nommus and the little man bounced backwards onto the pavement and into the gutter with me on top of him, tightly holding his hand. Big Foulkes was a bully and blamed the little man with loud abuse. Little Nommus got up bruised. He was shaking as we went to the High Street shops.

When we got back home I was giving Uncle Jim a jujube when I mentioned Big Foulkes knocking Nommus on the ground. Uncle said the man was just a big coward and he would put a stop to his gallop.

Big Foulkes swaggered into the Ship as the doors opened after we had all had our tea. Uncle Jim was just sprinkling some more sawdust over the floor. As Foulkes passed him, there was a sudden cafuffle. The publican grabbed the big man by the inside of his collar behind his head, and the big man turned like a top to face the door whence he had come.

The sinewy figure of Uncle Jim, former policeman in Egypt, former goldminer from Jo'burg, ran the hulk out through the door and applied a strong boot to his rear for good measure. That was the end of Big Foulkes, the bully. Strangely, after that it seemed to me that little Nommus used to stand still if we saw the big character and Foulkes avoided further encounters.

Old Sol was Old Maggie's brother, and he was a tinker, selling old and new tinned goods and bowls and so forth around the town. One day as he was working Holywell Road he saw me playing in the street. His big carthorse, with its cart behind, was moving along as he went from house to house plying his wares. Annie came out and took my hand to take me for a walk to the cob.

Old Sol said, "I'll give Jimmy a ride on Old Darby's back."

It was a big struggle because the horse was enormous and I was quite fearful of being perched upon the mountainous grey back.

I was not quite up when the horse shied and I fell off, down onto the road.

The great hind foot came down on my leg. I was picked up by a neighbor who ran out of her house to carry me to my Nain's. Nain bathed and bandaged my leg, but for at least forty years the weal showed that I had a small piece out of me and I can still put my finger on the spot offended some sixty-three years ago.

On a Sunday morning Ste would take me for a walk down to the Cob. To get there we would walk so far down Holywell Road, passing the red-bricked Courtaulds Building on the left. Then we turned right over a stile and usually crossed under works wagons standing on a siding, there being far too many to walk around.

We would cross main railway lines and also negotiate across some acid waste liquid running through channels from the works. Over the channels were little planks here and there.

That Sunday, as we were returning, Ste had crossed over at one point and I was shakily walking the plank to follow him.

"Give us yer 'and," said Ste, and as I did he yanked me towards him, but I slipped and was towed through the wicked water, all wet and tingly. It was another occasion for rapid action, and we sped back at a rate of knots to discard my saturated apparel and get me dropped into a tub with hot water and coal tar soap.

Let me tell just two more of my memories of the time I lived in the place of my birth, where I first saw the light of day as a baby "born in Flint without a shirt to his back," as the old saying used to go.

One day I was standing in my Taid's garden, complete with strong hobnailed laced-up boots, grey stockings, grey flannel suit and cap, when I did a little jig for Annie's benefit. As I did so, the sharp steel heel of my right boot kicked off a big wart about a quarter of an inch in diameter from my left knee. Instinctively I put the end of the big finger of my left hand on the spot and hobbled over to tell Nain.

"Quick, lad, off to Auntie Vi's. She's got some Friars' balsam to put a skin on it."

Thus informed, Annie and I were out of the house like a shot, and we made our way together through to the center of town, Trelawney Square, and on through Chester Road, then on a bit and left over the railway bridge, down to Henry Taylor Street and the abode of Auntie Vi and Uncle Walter Parry, the whole run being about a mile.

Auntie Vi duly applied the balsam, which worked like magic, and in my old age I see only a slight white blotch where, between us, Nain, Auntie Vi, Annie, and I cured the offending wart.

I will now sign off with the final event of my young five and a half years in this world, spent in Flint.

There was a primary infants' school in Chester Road, and at the age of five I was bound to attend some school or other. After all, the school inspector was none other than Molly Jones, Uncle Ike's brother and just the sort of man to report any delinquency in attendance.

So the picture remains with me of Ma and Auntie Hilda handing me over to a very tall grey-haired lady at the door of the red-brick school building. A few platitudes passed to assure confidence all round, and I was in the formidable situation of having to spend my first day in school. There was some plasticine, but there was also a lesson in arithmetic.

The infants' class was divided into halves, and somehow I found myself sitting by Georgie Hughes, who was working away at his figures. We both had sheets of paper with squares, and on the blackboard were what George called T. U. sums.

Well, the class teacher had told me to see if I could do the sums, though they were incomprehensible to me. The rest of the class had been attending school for some time, but my February birthday had thrown me out of phase.

George showed me what to do, and I duly did the same as he.

That day Georgie Hughes was my salvation. For a time, that is. He was also my downfall. Most of his sums were marked with an X, and as mine were exactly the same, I was treated to the same reward—a slight tap on the hand with a ruler.

On the way home Ma asked me how I'd got on. When I told her she said, "Well, you've learned something, Jimmy, never to crib in class," and after that I never did!

I can recall nothing more of my few weeks in Flint School, but at least I had broken the ice as far as academic experience was concerned.

Ever after Uncle Jim never let me forget the importance of scholarship. But we were not to stay with him long. My dad had not been sleeping, as we will see in the next episode. . . .

2. The Awakening

My dad was twenty-eight when he built the bungalow. The day came when we left the Ship with all our bits and pieces on an open model T Ford truck with Ma, Iris, and newly arrived baby Thirza tucked away with the driver. Dad and I were sitting on a sofa in the back with ropes entwined round the belongings, holding us all safely in place.

The fresh air of a spring morning in June 1922 saw me looking around as we trundled on our way from the Ship through Chester Road en route for County View, Woodlane, Hawarden, some eight miles away. I knew later that the route took us through Oakenholt, passing the paper mill, on our right, through Connah's Quay, and up to the crossroads at Queensferry, then turning right away from the river Dee and up the Mold Road as far as the Boar's Head, turning left through Ewloe village to the Springy and right again through the little village square, finally ascending the hill to our destination and new home.

The lorry stopped, then turned right into a side drive to the left of the bungalow as we faced it. Dad lifted me out, then helped Ma and the others out of the truck.

We all went to the front door, which opened with a Yale key into a cylinder lock. A wood paling fence separated the property from the hilly country road. This was creosoted and smelt fresh and antiseptic in the clean air.

Across the road was the wood—no houses. The top of the paled fence was artistic with inverted arcs, each eight feet long.

Inside the bungalow a passageway led to a square front room on the left and to a similar one on the right. A door from the passage also gave access to a big kitchen/living room with a Sefton type range on our right and at the other end of this living room were three doors, one straight on to a back kitchen, one in the left hand corner to another bedroom, and the third to a third bedroom, which was, so to speak, at the rear right hand corner of the bungalow.

Out through the back kitchen door we could see a view of a field and beyond it a farmhouse and more fields, and way over in the distance, Dad said, we could see the whole county and far away to the horizon, a tiny outline of the Liverpool Liver buildings twenty miles away. The distant view was always to conjure up my imagine of a strange land far away.

The layout of the property was completed by a washhouse and a water closet outside in the back, separated from the bungalow by a narrow passageway leading to the drive, on which now stood the Ford lorry.

As we bundled into the house and walked around in wonderment while the adults began to quickly bring in our belongings, Iris and I kept chatting to each other and getting in Ma's way as she busied herself with a fire to make up some tea and make the beds and so forth.

That first night Iris and I had to be silenced long after our usually early bedtime, as we kept up an unacceptable length of dialogue, but eventually it was all quiet and our new life in our own home had begun.

We were to spend the next four formative child years at the bungalow, with its hipped slate roof, its leaded light windows, its lovely plaster moldings to ceilings and cornices, its oil lamps, its quarry-tiled living room and wooden floor blocks to the sitting and bedrooms.

It was Percy Roberts whom Ma asked to collect me for my first day in school in the autumn term of 1922. Percy was a tall older boy of thirteen, who lived in a row of cottages about a quarter of a mile down the road on the same side as us.

"Is Jimmy ready?" he asked Ma.

"Hurry up, lad," she said as she packed me off.

Percy took me across the farmer's field over the ditch lying adjacent to the back of the bungalow. We ran the gauntlet to avoid the bull that Mr. Ted Weigh, the farmer, kept in the field. We called on Eddie, the farmer's son, and went on by the path that took us past the old, disused brick kilns and "crown hills" of the old waste tips from the one time brick works. We continued past a plantation of fir trees before eventually getting to Ewloe Green Council School, some one-and-a-quarter-mile distance from the place where I lived.

The event was somehow far more amenable than my introduction to the town infants' school at Flint.

As a new pupil expected by Miss Bennett, I was ushered into the infants' end of the red-brick building of the school, all on one floor.

There were cards with drawings and numbers on the walls, and the infants' pupils sat in pairs. Half were second class and half first class. When asked, "Have you been to school before?" I said, "Yes, miss," which was true, but only just, and I was put into first class. There was modeling with plasticine and the alphabet and counting and simple times tables, all of which I struggled through without chastisement. Everyone was ever so friendly in this little village school, and we had playtime in the morning and afternoon.

Scholarship and senior pupils and headmaster, Ewloe Green Council School. The author is standing on the extreme right. Somewhat disheveled, he had just arrived to join the group, impatiently waiting during the lunch hour.

At lunchtime I had sandwiches from home. Ma had added a bottle of tea to wash them down. In the years that followed, I was to get used to the sandwiches, so often bacon but sometimes bananas or jam with a piece of cake. But often enough when I got a bit older I would run home for lunch and then run back again during the one-and-a-half-hour midday break.

In the four years I attended the elementary school at Ewloe Green, my life assumed a stable pattern of learning my lessons, complemented by a wide scope of new experiences to do with life at home and at play.

At school I took pride in usually being in the forefront of my classes, automatically, it seemed, sticking up my hand sharply to get Teacher's attention to answer the questions as they were put to the class at large, whether in dictation, sums, or general knowledge, and so, despite the long walks or running to school in all kinds of weather, I was able to progress without hindrance and, on the whole, as one of the favorite pupils. In a nutshell, while I went from $5\frac{1}{2}$ to $9\frac{1}{2}$ years of age when living at the bungalow, my school cap size grew from $6\frac{1}{4}$ to $6\frac{3}{4}$ and I went inexorably toward my prescholarship year from first class infants to standard 1, standard 2, jumping standard 3 to complete standard 4 at the summer term of 1926, when we would move to live quite close to the school, as I will recall later.

In such a school in those days there were sliding partitions and they were so thin one could hear other classes chanting spellings or answering questions. It was a real hive of activity inside the school and the sort of place that stimulated a pupil who was acutely interested in the competition of its goings on. It was as though those with a light heart were floated along on a floodtide while those with a heavy heart tended to sink and get left behind in the mud.

A little chord in my memory is when in standard 2 four of us at the back of the class were busy with a little composition on the subject of "washing day." Now I did not know how to start. After all, we were always in school on washing day and it was not something I had particularly noticed. So I glanced to my left and saw that Clifford Evans had already written: "Mother washes on a Monday." Then I glanced to my right and saw that John Evans had written: "My mother washes on a Tuesday."

Now I did not want it to be obvious that I had had the inspiration for my first sentence from either of my neighbours, so I wrote: "My mother washes on a Thursday." I cannot remember what else went into the composition, but I can remember that my mother gave me a good telling off when, as usual, I related the events of the day to her and she learned I had misled all to suppose she was a person who washed on a

Thursday. Anyway, the next day I owned up to Miss Garratt, who seemed to think the whole thing rather amusing.

I suppose there was always an accommodation shortage in schools. We used to go into the corridor for spelling sessions, and that meant perhaps twenty of us standing around in a big arc, a few in the middle sitting on a form and, as it were, the head and tail of the arc shaped by those standing in the single file of the class.

On a certain day out came a big word for the class to spell with the carrot of going home early: *entertainment*. I stuck up my hand. Miss Garratt, a fortyish, dark-haired, slightly grey, semi serious type of teacher, picked on me, and I got it right.

"You may go, James," she said.

Percy Jacques slid off the end of the form to allow me to pass, but not before I had slipped under the form to leave it in front of my own seat position. As I went by him he whispered, "I'll get you," and stroked the buttons on my single-breasted grey jacket.

I thought nothing more about it but rushed home (the journey seemed only a few hundred yards) in my elation to report to my mother the big success of the day.

"You've done well, son," she said.

Her sister, my Auntie Annie, who was only five years older than myself, had been staying over with us for a few days. "I don't think I could spell it myself, Jim," she said.

After tea I went down the lane to the square to play marbles with the lads under the lamp. Percy came up to me. He was incensed that I had apparently ignored his friendly gesture to let me off the form seat when he had graciously stood up.

I was soon defending myself, as he aimed a blow at my face. I grabbed him round the small of the back in a child's idea of a bear hug, my favorite grip in such a situation. We both fell to the ground or, rather, the hard road, with me on top of Percy, so I gave his head a little tap on the road for good measure, whereupon he yelped and I got off to let him free, my own bruise just a scratch behind my wrist.

It was the event of the evening, and the marble playing was forgotten, as we all quickly dispersed for home. Going up the dark lane on my return journey, I did not think of my tousled condition until I got into the house, where Ma was waiting for me to send me off to bed, with Iris already asleep.

"What on earth have you been doing?" she asked me.

"Fighting, Ma. Percy Jacques came at me because I did not get off the end of the form when I spelt *entertainment* in school today."

"It looks as though he gave you a good pasting," she said, whereupon the conflict caused me to burst into tears, as I burst out, "I *beat* him."

Conker time is important in a boy's life. I had just had my conker niner broken in school by Dennis Griffiths when I took out my slating nail to bore a hole into a new conker, one of many in my pocket. As I pushed and bore the sharp end of the galvanised round-headed nail into the horse chestnut seed, it suddenly plunged into my left hand, ably helped by a big push on the head of the nail by my own right hand. It was extremely painful as it went in nigh up to the back of my grubby little paw.

It was fortuitous that in the school that day was a nurse who was doing her rounds of the schools, looking for ringworms and so forth, and as I sped into the school from the playground I ran right into her with the boys chorusing my misfortune.

The young lady, for such she was, rushed me quickly to a washbasin and plunged my hand into the hot water as hot as she could bear, which was a jolly sight hotter than my limit. She then squeezed my hand till the blood squirted out into the water and finally put some paste and bandaged it up. There were no late ill effects, so I was lucky that she was on hand to attend to a nasty situation.

Not so lucky was Harry Bryan on a winter's day when the boys were all sliding on the Flash. This was a frozen over area of low-lying field adjacent to the school. On the day in question, Harry, one of the big boys of standard 6, had been excelling when the ice broke.

In went Harry. When we saw him coming into school his face was a mess of blood. He was a brave lad and made no complaint but vainly tried to stop the blood, washing himself in a washbowl and applying his hanky. There was no one on hand to attend to it properly. It seems that he received medical attention far too late and, in later years, his nose was foreshortened and disfigured as a result of the most unfortunate accident.

Do not think that going to school and back home again was a mere routine matter, though some things became a sort of ritual.

There were several stiles on the way, and these had to be vaulted. There was a favourite place where we used to leap over a ditch of varying widths, and so long-jump practice was possible, with the penalty of getting wet if we did not make it at the spot we chose for ourselves. There was a favourite tree with branches to drop off of according to the prowess of each individual.

There was also the halfway point, old tips of waste material we knew as "crown hills," with little craters we could sit in to shelter awhile from the biting winter wind. I used to carry a piece of candle in my pocket and a box of Pioneer matches so as to warm my hands, which I do recall had often got chilblains in the winter.

In the deep snow one's Wellingtons were invaluable, but the journey seemed five times as far.

So much for this early school period. What was happening at that time on the leisure front at home? How about a few glimpses as I fill in the picture of those postwar years in the early twenties?

What was Dad doing? Well, he was earning his living by building more bungalows, in addition to doing various jobs of alterations and extensions for people. Like for instance, going up the Woodlane on the same right hand side as us, he built similar bungalows for my Auntie Thirza, his eldest sister, and Uncle Fred Sheppard, her husband, for Johnny Tomlinson, and at the top for Mr. and Mrs. George Clarke, a retired couple who kept chickens.

Now Uncle Fred was a Lancashire man, and after Dad bought his first Ford lorry—a 30 hundredweight flat-bottomed truck, Uncle Fred decided to do the same in order to increase his round selling earthenware goods made in the local pottery at Buckley.

Previously Uncle Fred had a horse and cart and in his Lancashire brogue would call out, "Butley mugs, Butley mugs," but then the horse was so sprightly the cart did not wait long enough for potential customers to come out of their houses.

Uncle Fred was proud of his enterprising attitude towards things. Thus he learned to drive his lorry in half an hour, but he was very overbearing with my Auntie Thirza, doing the food *his* way, and in other things, too, she was left with little to do, as she had no children, and even at my age I could see she was awfully timid when he was around.

Johnny Tomlinson had a smallholding behind his bungalow, with pigs and chickens and a couple of cows. To quote him, when he saw me with some loose change in my hands, he said, "Thou'll buy a 'poss' when thou loses the value of one."

George Clarke had a son who was a civil engineer working in South America. We, Iris and I, often used to visit Aunt Emily Clarke and have a piece of cake.

Farther down the road there were a couple more bungalows Dad built, one for Sam Whitehead and one for the Steel family. Dad was ably helped in all these by Ma's brother Ste, in his teens, who did the carpentry under Dad's guidance.

Alf Tudor, the bricklayer, did not waste time in getting up the brickwork. Labouring help came from the immediate area. Jim Millington drove our first Ford truck, registration no. FR5308.

Now when I was just over nine years old I well remember Jim Millington giving me a tip about driving as we came from Ewloe village to turn into a very sharp right hand turn into the Springy Hill at the bottom of Woodlane.

"Say to yourself, 'I'm *going* to get around this corner,' " was Jim's prognostication. One day he let me hold the wheel and I was *almost* able to put it into practice, but at the last moment it was clear I needed stronger hands, and Jim came to the rescue with a firm yank at the steering wheel to ensure we did make it safely around the corner.

A favourite hobby of mine was making bows and arrows. I would cut off a hazel sapling and peel off the bark. Then I would fashion a few arrows, also debarked and pointed, with a nick to engage the string of the bow, itself securely tied at both ends. In time it was surprising how often I was able to hit a ready-made bull's-eye in the form of a one-inch-diameter hole in the iron post of the clothesline.

Another typical boy's pastime was the catapult made from a stout Y-shaped piece of wood with square sectioned rubbers and a pouch made from the tongue of an old boot.

One device I contrived unique to me was a wheeled cart, the chassis of an old pram, duly modified and steered with a rudder at the back like a dinghy, the rudder being a heavy iron bar trailing on the ground as I sped down the hill at a rate of knots. Unfortunately, this activity was to lead to disaster one day, not for me, but for a man who used to go down the hill passing our home on his bicycle to Summer's Ironworks at Queensferry, about three miles away.

I was pushing the *Rover,* my wheeled cart, out onto the road when the unfortunate cyclist ran into my chariot. The bike went careering down the hill with a heck of a wobble, while the man struggled both to keep his balance and to dismount. He had to leave his machine at our place and run to work. That day he was sent home from work as a penalty for being late. He spent the day straightening up his wheel and, fortunately, was able to go to work the following day. After that, I was more careful with my own vehicle.

In the early twenties it was quite common to see the odd tramp trudging along. Some were regulars, who spent the night in the local brickworks. There came the Saturday morning when one such individual came round to the back door to ask for something to eat. Usually the request from him and others like him would be for a drink of water or a tin of tea from the pot, plus a crust of bread with some jam on it. This day, the tramp espied a quarter of a fruitcake my mother had on the back kitchen table and demanded he be given the delicacy. "That'll do, missus," he bawled out in a voice to intimidate the Devil himself. My mother was scared into letting him have it, and he made off down the lane. Dad came into the house a couple of minutes or so later and Ma told him, whereupon down the hill, on his bike, he espied the miscreant walking, still wolfing the last bite of cake.

Dad put out his right hand and belted the old codger behind the earhole and knocked him flat on his face. Then he got off his bike, and as the old rogue got up he booted him in the behind. The wicked oaf never called again at the bungalow, though others did as usual.

About this time I was visiting my Nain in Flint for my Easter holidays when Taid brought home a secondhand Raleigh bicycle. "I got it from Lord Mostyn's home. It used to belong to his son," said my Taid, the captain of the *Pilot*, a coasting sailing vessel of some 103 tons, which was furnished with an auxiliary Perkins semidiesel engine and which used to put in to Mostyn with iron ore. We straightway went through the narrow wooden gate beside the midden at the bottom of the long, narrow garden and out into the field used by the Flint town football club, with the dominant sight of the gasometer in one corner. The lingering odour of a gasworks stays in the lungs like the smell of a sweet tobacco to a smoker. It has a taste of the essential ingredient in the air one breathes; in other words, it grows on you. I can still smell it.

Taid was about to teach me to ride. I was just able to reach the pedals when he took off the seat and substituted a piece of cloth. His method was to hold me back as I stood on the pedals trying to make him release his grip. Then he let me go off like the stone from a catapult. Round and round I went in ever decreasing circles, frantically trying to miss the one obstacle on the whole field of bare, dirty grass—a half brick. In the end I ran, petrified, straight into the brick and came off over the handlebars. *C'est la vie,* I might have understood in later years.

There were a multitude of little events of this era of my life, like riding on the shining motorbike behind Uncle Tommy, a Sunbeam that had a twist grip and ran as quietly as a Rolls, and there was the Airedale dog we kept in a wooden enclosure. Jack had a bark like a lion and was really a fierce yard dog, yet instinctively did not bark when a relative came to the house, even if it were the first visit. There was the time I fell into the pit in the wood and kept company with a snake until Iris managed to get help two hours later, when Dad came home. Then there was the time Auntie Doris turned up with her new boyfriend, Fred Chesters, and persuaded him to jump over the ditch and hedge, only to see the poor chap break his ankle. There was blackberrying and falling into the spiky brambles while reaching for the biggest, juiciest-looking one at the top of the bush. There were the great annual events in those parts—the Buckley Jubilee, the Whit Monday sports at Hawarden Park, and the Whit Tuesday sports at the Flint fields.

Of the other 990 or so other incidents that titillate my fancy I'll just mention one significant occurrence about early 1926, the year of the general strike, when we were left with no coal.

As it happened, our bungalow was sited on an outcrop of coal. Dad was quick to realize it. It was of poor quality, but so near the surface. It came under the bungalow sloping up into our garden down the slope of the ground parallel to the main road.

So Dad called in an old miner friend of my grandad's, and together they sunk a small shaft into the coaly ground.

What a fine adventure for a boy of nine to see the operation of a tiny coal mine opening out in the very garden in front of him as the weeks rolled by.

First of all, the idea was just to let us have coal to burn in our grate. Then, as the stock came out of the earth and local people collected barrow loads to keep themselves going, Dad managed to spare some for coal suppliers in the Wirral, who were so pleased to do something for their customers. The going price was three shillings and sixpence per hundredweight bag.

It was down the pit I first learned the value of maths. The old miner was good at it and pulled out of his pocket a new halfpenny to demonstrate as we were sharing a can of tea.

"Look at this, Jimmy. What is its diameter?"

He pulled out of his pocket a two-foot rule, and we measured across. "One inch," we concluded, in harmony.

"What is the circumference?" asked Joe Williams, and the forefinger of his gnarled hand delicately traced the edge.

I looked at him blankly before he came out with one fit for Archimedes. "It is one inch multiplied by the value of pi," he said. Then he marked the edge of the halfpenny and rolled it from the end of the rule to just past the three-inch mark to stop the coin where it had made a complete turn. "About three and a seventh," said old Joe.

That was the first demonstration I had showing the relationship between a straight line and a circle, and it says much for its value to me that I never forgot it, and it is as good a note as any to finish on as we pass to the next episode of my life. . . .

3. Homework and All That

One of the multitude of things that were happening but which I did not find space to mention in chapter 2 was the building of "The Houses."

When we lived at County View, while I was fully amused, entertained, and preoccupied in my own little world, waking up to all the wonders around me, Dad was working as only a man who works for himself can work. In between doing work on buildings and houses for other people within a radius of about ten miles, he was also finding time to erect a pair of houses on a plot of land opposite the Ewloe Green Council School and on the corner of Green Lane's junction with Mold Road. We were to live in one of the pair.

The plot cost £145. I had been able to follow the progress of the houses' construction by nipping across the road from school before going home and having a chat with the two Alf Tudors, one senior and one junior Alf, who did the bricklaying, with Uncle Ste doing the carpentry, old Josh Edwards the painting, and Hughie Finch the driver of our more recent Ford Truck, DM7225. The building of the house, with its trenches, its lime pit, its doors and windows and slated roof and plastered walls, all served to educate me to construction work at a very young age.

It must have been about our summer holidays when Dad let the bungalow to a tenant and took us over to live in Hendre Villa, not quite finished, as on arrival we had for company a living room stacked up with cement and nails and all sorts.

Still, the plot belonging to our house, with its builder's yard area right opposite the school, had obvious advantages for all of us and there was a general atmosphere of expectancy. Certainly no complaints were in evidence.

The old lime pit still had some slaked lime in it, so this was railed off in case we children were too venturesome. By this time our family included baby Horace, born on July 29, 1925.

It was with this new beginning that Dad began to establish himself as a builder and contractor and simultaneously I began to do homework for the scholarship. At this time there was a great rapport between father and son. Indeed, all the family were so much occupied that life was never dull. We seemed to believe that the world revolved around our home, with so much going on and everyone from the village passing by with a cheery word over the hedge or popping in for a chat or possibly a look around the yard or both.

Let me tell you about just a few little events that illustrate the course of life in Ewloe Green between the years 1926 and 1935, years in which I took the scholarship examination. I attended Ewloe Green Council School in 1927 and then attended the Hawarden County Grammar School until 1934, followed by an eight-month period working for my father on a housing estate in Connah's Quay prior to a fortunate turn of events that precipitated my entry into the University of Liverpool to study for an engineering degree.

As I was the eldest child of a family of five offspring (Bruce was born on May 17, 1932), it was important to my parents that I should do well, so I was given every encouragement at school.

I am of the opinion, however, that my success in gaining full scholarship to the secondary school at Hawarden was due more to my innate desire to rush through homework in order to go out to play football or cricket in the evenings. This probably speeded up my thinking.

Of course, whenever he paid us a visit, Uncle Jim exhorted me to pursue my academic work with vigour, but probably the biggest single factor leading to my being in general family favour after the June exams of 1927 was the physical bouyancy I had gained in playing football in the open field that lay across an open ditch still uncovered to this day. On that field, belonging to Farmer Jones, who lived along the Green Lane, my village pals and I had many a vigorous game till we could not see the ball because of the twilight merging into darkness and this come rain, come shine.

At such a time I would arrive in the house, stockings down to my ankles, quickly wipe my hands in the sink, and go into the front room to saw away at my Maidstone violin under the command of my private music teacher, Miss Millicent Booth. Millie did not deserve me. She was a delightful young lady in her early twenties, and with each exercise completed she would treat me to quite a professional rendering of "Tiptoe through the Tulips" or an equally suitable tune.

I remember I got as far as exercise 19 in the ten lessons of the quarter. Then it was more than my mother could bear, and from then on my musical endeavours were limited to try my hand on the piano after my young sisters, Iris and Thirza, had completed their own practices. I was to have no more music tuition after the collapse of my father's aspirations for me to shine on the very violin in which he himself had an equal lack of success. 'Nough said.

In the summer we boys were caught up with cricket on a field provided by Tom Evans, John Evans's dad, and used equipment provided by Neville Roberts, who used to remove the means of play as soon as his parents, usually his mother, called him in for his tea or to do his homework.

Apropos of this, there was a short outburst of field violence involving little Jim Tellett, who was always telling us he had been a weakling at birth. "I had a band around me when I was small," he would say.

Neville was responding to the parental call, picking up his stumps, when the small one was moved to chant a little monologue.

"My bat, my ball, my wickets, tara," proclaimed the frail one, whereupon the incensed Neville grabbed the bat held by little Jim and thumped the budding bard with the handle of the three-springer in the manner of a battering ram, which had the effect of felling the youngster to the ground with a mournful cry of, "Oh, me, oh, me, o-o-o-h."

All we boys, John Evans, Ron Weigh, John Rowlands, and the rest, dashed around Neville, who was administered summary chastisement, little Jim soon bouncing up and down, enjoying the turn of events, with Neville in chagrin before he escaped from the field.

Under the chapel lamp we lads would often congregate to play marbles. In a later chapter I will have a word to say on a certain spectator interested in this activity.

Although my gravitation from the elementary to the secondary school did not mean I had to leave the house in the village of Ewloe Green, it did mean the loosening of bonds with friends who were not to come along to the new school. They were either to go to the new central school at Queensferry or remain to finish at the village school, working up to standards 6 and 7, and even ex-7 if they had the time and inclination, before leaving school at fourteen years of age.

The night before I sat the scholarship, later called the eleven plus, though at the time I was eleven minus eight months, my dad had me doing some figuring to estimate the cost of some works he was tendering for in the Flintshire area. At the time I thought I was being rather put in a state of stress, as I also had to find time to revise my lessons in arithmetic, algebra, and English, but I expect the old man's idea was to take my mind off the tension of my situation. At any rate, I was none the worse for my overtime on Dad's behalf.

For his part, he was doing various speculative pairs of houses and alteration jobs and began to take an interest in sewerage work. An older brother, Ernest, was able to discuss with him the ins and outs of how they should go about tackling a small scheme for the Hawarden RD Council entailing some gravity flow, a pumping main, and a railway crossing.

So now Dad was thinking of himself as H. Price, builder and sewerage contractor. He had also been on with some houses at Mount Pleasant, Flint, near the Mill Tavern public house.

Riding into the county school in September 1927, my first encounter with the establishment was to hear a cry from the caretaker and gardener

Old Man Jones: "Ye boy, get off that bicycle!" In the years to come Man Jones was to receive quite a bit of help from the boys in gardening hours when they did little else but turn over the soil for him to plant vegetables, largely to be used in "school fourpennies"—the price of five school dinners was two shillings, which being divided by five equaled 4.8 pence, hence the name for the dinners.

As we dug in deep during gardening sessions, Jones would come along and stand at the end of a row and call out, "Ye boy, bury ye green," to ensure all grass and other verdure was duly turned in to rot as manure.

Now Mrs. Jones was the cook in this coeducational school. When I say "coed," I mean there were the boys and there were the girls. In theory we mixed freely; in practice the natural choice of subject—girls usually went for literature and art classes, boys for maths and science—made for a dividing up of the school into male and female classes.

The usual sort of division of forms in any year would be, say, IIIB 95 percent boys, IIIG 95 percent girls, and IIIC, commercial class, with some 80 percent girls and 20 percent boys. The forms went from second to sixth. The sixth forms were lower and upper, with a third-year sixth for those taking university scholarship about the age of eighteen.

The school head was Arthur Lyon, M.A., who used to roar like a lion and lay about recalcitrant boys with a stick, stiff and thick. Fortunately, I managed to keep out of his way for my first two years and he retired at the end of the school year in 1929. B.M. (Ben) Jones, M.A., B.Sc., took his place. He came from Cardiff and introduced rugger to the school. Prior to this, the school's great pride in sport was its soccer team, better than most county schools, or so we claimed. The girls played hockey and they too were pretty good.

I am deeply grateful for the academic and extracurricular opportunities of Hawarden County School. When I began, I was ten and a half years old, a mere four-foot, four-and-a-half-inch, five-stoner, and at the end of the first term and exam work I found myself eleventh in a class of thirty-five pupils.

Seven years later, at the end of the summer term of 1934, I was possessed of nine subjects at Central Welsh Board matriculation level and also had obtained physics and pure and applied maths at higher level, being also fifteen inches and four stone bigger.

Soccer, cricket, and fives were my favourite participatory sports.

School dinners were a communal affair. For each table boys attending (some took sandwiches) would sit in order of seniority. At my table in my first year, a teacher headed the table with older big boys—including prefects—on either side of him and progressively younger boys down to young Price at the very bottom of the table.

It was Tuesday and we had just had shepherd's pie when it came to rice pudding, and Les Mothersole beckoned me up to him. "Go and ask Ma Jones for the sugar," he ordered. I had to go. You don't argue with so much avoirdupois from the lower sixth. Over in the kitchen I put up my hand as I stood by Ma Jones, the cook wife of Old Man Jones, the gardener.

Ma Davies, the headmistress, was chatting to the cook. "Who asked you for that?" she inquired.

"It was Mothersole, miss," said I.

"Up to their old games," she said to Ma Jones. To me the great lady added, "Tell Mothersole Miss Davies says he must come himself for the sugar."

When I returned there was much tittering at the table. This was a form of humour new to me. Meantime the pudding had been shared out and it was interesting to note that the size of the shares was in proportion to the square of the ages of the pupils.

Ironically, by the time I had gravitated up to the sixth form and sat near the head of the table, democracy had established itself and all had equal shares, which means I lost out on grounds of equity, taking my whole period in the school into account.

One of the boys at the top of the table was Kip Walton. His father had a jewelry shop. One day Kip was absent from school. Word came that there had been a fire and his father was badly burned. Kip went into hospital with him and had large areas of his own skin stripped off his body to save the life of his father. Thereafter he was a hero, greatly revered by every boy in the school.

When Shotton Ironworks had to shed labour due to the national depression, there was the story of Cyril Courtman's dad. The boy who put it about was "Span" (Aubrey) Deakin, son of the pioneering trade union general secretary Arthur Deakin. It seems that Mr. Courtman was one of the underbosses in the administration side of Dick Summers's Ironworks. He was approached by a man in top management and asked to produce a list of those who could be dispensed with. Old Courtman was unable to put down one name, so he put down his own first, then, with a better conscience, added a few more whom he thought would be least hurt by the dismissal notices.

Music recitals in the assembly hall came now and again. Familiar was the recital of the Bangor Trio playing the "Gipsy Rondo" and other works. Before they started, the head always came out with a "We fry ours in lard" joke, just to ensure we did not talk during the performance.

Church attendance for me was regular. My mother used to see to it that I went to Hawarden Parish Church, a good one and a half miles from our home in Ewloe Green.

I shall now acquaint you with one small hiccough in the smooth running of my betterfication. At the age of twelve, in 1929, I began to be exercised by the fact that my hair would not lie down along the parting; it was always sticking up in the manner of "sore fingers." I took it into my head to do something about it.

"Ma," I said, "I think I'll have my head shaved and train my hair from scratch. I've heard it can be done. I'm fed up with trying to plaster my hair down."

She raised her eyebrows but could not deny I had a problem. "Up to you, lad," she said, "if you've made up your mind."

So Saturday found me sitting in the chair with Dick Davies, who could have doubled for Ray Reardon, the snooker professional, looking down at the sight of a head resembling a hedgehog's back below him and asking the inevitable question.

"Back and sides as usual, Jimmy?"

"No, Dick, off with the lot," was my reply.

"You don't mean?"

"I do. A Gandhi crop."

"But it hasn't started around here yet. You'll get an awful drubbing from everyone," went on Dick, the specialist.

"What's the difference? I do now with my hair uncontrollable. I want to train it from scratch."

"Well, reluctantly, young man, reluctantly. Does your mother agree to it?"

"Well, I did mention it in passing."

"In passing what?"

"In passing out of the house. Dick, please, get on with it."

He started at the front with the cutting shears, and took it all off just like a man shearing a sheep. There was no adjustment in price up or down. I paid him the usual fourpence, promising to see him again in another month. "If my hair grows again," was my Parthian shot.

There were a few wisecracks on arriving home. Hughie Finch, our Ford lorry driver, said after taking a good look, "Thee 'ead's as bald as a billiard ball 'cepting tha's gotta square 'ead."

Grandad, Dad's dad, coming into the yard as he was wont to, leaning slightly forward with old grey trilby stuck on like glue and sculling himself along with his walking stick, asked me a wounding question: "Who's shobbinned thee 'air?"

Ma made no adverse comment. "So you've done it, lad," was all she'd say.

The next morning I stayed in bed reading Robert Louis Stevenson. Ma came to root me out about 9:45 A.M.

"What about church?" she asked.

"I can't go without hair, Ma."

"No excuse. You're being confirmed soon."

I had no further argument. I went.

Ron Weigh and I were just in time to see Lawson, the form and lit master, sitting in his pew halfway down the aisle of Hawarden Parish Church. The only empty seats were in the pew in front of him. We went as invisibly as possible to our ordained seats, and listening intently to the rector, I kept my head looking directly at the man in cloth like a blinkered horse on a busy street, hoping Lawson would not recognize me from behind.

The next day, the Monday, in literature class the lit wit said, "We'll have the hairless wonder."

I stood up while Pug Jamer and Bill Cox, two class heavies, bore into my head on either side with their organ stop eyeballs.

I began:

" 'Give to me the life I love,
Let the lave go by me.' "

The lit wit couldn't resist a wisecrack at this point and interposed, "Yes, very apt in your case, I must say."

Jamer and Cox guffawed, followed by the rest of the class, but the rumpus was quelled quickly by Mr. E. G. Lawson, M.A., who withered them all with one stern, sweeping glance.

I felt myself colour up, then dug in as I spouted out, firmly and resolutely, the rest of the piece:

" 'Give the jolly heavens above,
And the byway nigh me,
Bed in the bush, with stars to see,
Bread I dip in the river,
There's the life for a man like me,
There's the life forever.' "

The form master said, "Sit down, Price. At least you've done your homework."

It is interesting now to reflect from such a distance in time that boys began to have Gandhi crops very quickly after that—Pug Jamer (Leslie James), Bill Cox (William Albert Clive Cox), Hughie Finch, John Evans, my mate in the village, then most of our class till one day I confessed to Ma my sympathy for Mr. Lawson's problem of disciplining as each extra bare head asserted its membership to the new fashion.

Suffice it to say that my hair never gave me any trouble after that. Eventually it became too tired to stick up, and since the age of thirty-five or so I have not had much on top to lie flat, my parting being about four inches wide.

Apropos of confirmation, my church pal Ron Weigh and I were confirmed in 1932 at fifteen years of age. Father Roe was the priest who prepared us for the ceremony, which was carried out by the bishop of St. Asaph about Maytime. After that we took our procedures seriously, duly going without food or drink each Sunday before taking Holy Communion.

One Sunday, it was such a lovely day Ron and I decided to take a run on our bikes as far as Mold, some four miles away. Actually, Ron borrowed my Uncle Ike's bike, as there was something wrong with his own. Ron had to stretch a bit to reach the pedals. He tended to sit on the crossbar of the machine.

Arrived at Mold, we decided to press on to Ruthin, and the bug having bitten us, we went on to Denbigh, finally making a detour back through Rhuddlan and back home, a circular trip in excess of forty miles, which occupied us from 11:00 A.M. till 3:00 P.M., sustained only by a drink of water at a pump in addition to the Communion wine and wafer. On arrival we were shattered.

"Where on earth have you been?" Ma had worried herself sick, keeping in the oven my dinner of new minted potatoes, green peas, tender lamb, and a big helping of an apple suet pudding with custard. Apart from the custard, I was too weak to eat it. However, after an hour's sleep, the contents of the platters soon disappeared as a much relished Sunday tea.

What about the goings on of the builders' yard? Well, the place offered a fund of knowledge and experience for a young teenage boy to enjoy as he grew up. Dad had rebuilt, in the place of a corrugated iron timber shed that blew away, a substantial brick-and-slate two-storey building, known as the workshop. At ground level there was provision at one end for two vehicles to be garaged and at the other end was a single door into a heavy workshop section with a very solid bench, a peg vice, and lots of pigeon holes in a wall cabinet for easy access to nails, screws, plumbing fittings and, leading off at right angles to the main building, a single-storey shed with a circular saw and bench for reduction of large timbers to handy sizes. These were transferred upstairs to be made into door and window frames. I was often drawn in by the joiner, Ste Bithell, Ma's brother, to help him with mortising out timber sections for joints. Ste was a good joiner. The planing machine was also upstairs

25

and fairly zipped through the work of rebating and moulding the square section timbers into the necessary shapes of the units required.

When times were slack about 1931, Dad kept the loyal men close to him simply making more doors and windows so they would not have to go on the dole.

From time to time, during the lunch break Uncle Ste would place on the saw bench upstairs an old quarter-size Riley billiard table he had salvaged from the old Ship Inn when we left Flint in 1922. It was a bit worse for wear with the bottom cushion rail rescrewed on and the rubber cushions all hard; there was a red composition ball and two yellow coloured ivory balls serving as the whites. Nevertheless, the old table gave me an extreme amount of pleasure and often on a wet day when homework palled I was able to score an occasional hundred break, enough to give any youngster a lift.

Dad's business was expanding and into the yard came an assortment of items. There was a quadrant crane for loading and unloading, a pan mixer for black mortar, a Graham Dodge, lots of spare pipes, surplus from sewerage works, and one item of special interest and great character; a ten-ton flat-bottomed Foden steam engine truck, registration number M7, made in 1905. Fred Edwards used to drive it. It would go all day on a hundredweight of coal but only did about five miles per hour. The driver was paid £3.12s .0d per week—a good wage in those days. When the vehicle went down the main road, the pattern of the tread of its solid iron wheels was often impressed on the soft surfaces of tar and chippings typically to be seen around the land. We would all go out into the road to see Fred disappear from view, not expecting him back till the next day, as his trip would be for about fifty miles or so on the North Wales coast. If he ran out of water, he could fill up at a roadside ditch. If he ran out of coal, he could burn some wood from dead branches, more available at the roadside than nowadays. When the new Sentinels came in with solid rubber tyres to their wheels, Dad must have realized that his old Foden could no longer compete in haulage, so he had it cut down for scrap using acetyline burner and cutting tackle. Over half a century later, I can still remember the sadness I felt to see old Fred cutting up his own vehicle, taking out the tubes of the boiler, and cutting it all up into pieces until finally along came the scrap merchants to take it away. When, in later years, after supervising the M1 major motorway, the pioneering work in chapter 14, "The Great Wide Way," I approached the Cheshire licencing authorities for the release of the number M7, I was told it was "not available." This was a pity, as the number would have meant a lot in sentimental value as a link between my father's work and my own, thirty years on. C'est la vie. . . . (It seems that my own vivid "M7" memory of the registration number was incomplete. Recently, National Traction Engine Trust records elicited that the number was M761.)

One day I went up with Hughie to collect an old Dennis chassis that was available at a farm at Ewloe Barn, a neighbour of my grandad's farm, Smithy Farm, and some mile and a half from Hendre Villa, our home. This meant that I, a boy of fourteen, had to stand on the cruciform of the bare steel chassis and steer the skeleton of a vehicle with no body, not to say seats, and use my dexterity to apply the foot brake to keep the tow rope taut all the way to the yard. Thus arrived, the next few weeks saw Ste building a caravan superstructure on the old Dennis undercarriage. This bungalow-shaped caravan (it had two bay windows to the central living room and two bunk compartments with sliding doors at either end) was destined by Dad to be sited in a seaside holiday campsite, Terfyn Pella, near Prestatyn, near Rhyl, on the Flintshire coast.

As it trundled off behind the fifty-hundredweight Dodge tipper lorry, before it got to the Boar's Head the solid rubber type spoked wheels were seen to develop an unacceptable wobble.

The practical minds of Dad, Uncle Ste, Hughie, and Bill Gittens were confounded. The young theoretician had a suggestion. Tentatively, and with appropriate delicacy, I explained my view to my dad. The towing frame made by the blacksmith was a simple V-shaped affair made of one-inch reinforcing bar. In order to ensure flexibility, each leg of the bar had been shaped with a graceful curve. It was this flexibility that caused the wobbling, as the front axle shook itself out of control as it bumped along over the potholes. Therefore, the frame had to be rigid and with a strut cross for extra rigidity it would then be an A-frame. Dad took the point. The towing frame was sent back to the blacksmith in Liverpool Road; Buckley and the big van went off to its destination without further ado. Dad asked me to go along with the Dodge towing the van. My stock had gone up. My education was paying off. The caravan was to be the family holiday home for many a year on a fixed site at the Terfyn Pella caravan site.

Let me try to capture the flavour of these years leading up to the end of my time at Hawarden School in June 1934 by relating briefly a few of my duties and activities in the next few paragraphs.

The hedge around our house and yard was a lengthy one, mixed but mostly of hawthorn. It was on this that I began to build up my shoulder muscles and biceps. When Ma and Dad went for their regular Saturday walk in the summer months, from the age of eleven I used to earn a half a crown by trimming the hedge, taking pride in keeping it to a good line.

Jim, when a scholar at Hawarden County, paid several visits to his birthplace, 39 Holywell Road, Flint. At age eleven in 1928. At age thirteen, in 1930, with cousins Brenda and Sheila Parry. At age seventeen, in 1934, as a sixth-former.

The bicycle I bought from my pocket money to replace the old, well-worn version was a splendid three-speed Raleigh Sports, which had inspired me to ride off into the sun after Holy Communion one Sunday. The bicycle for a boy can be multifunctional. For me it was the means of going off camping to Rhuddlan, Prestatyn, and Towyn, near Rhyl, some twenty-eight miles away. It propelled me around the whole district in Flintshire. It was my faithful ally and friend when I went to collect rents from the tiny cottages owned by my Uncle Jim in Connah's Quay and by my grandad in Church Road, Buckley. It was instrumental in letting me visit Auntie Nell every Monday on my round after school to collect rents, as she slowly slipped into oblivion after lying for weeks running into months with her terminal illness, a sadly profound experience.

Speaking of multifunctional, I myself was a bit of a jack-of-all-trades in those days. You see, Dad was building houses and doing sewerage schemes in Flint, Connah's Quay, Hawarden, Caergwrle, and Congleton. I helped along by dropping in as office boy, driver's mate, measuring up with Dad's agent, Fred Robbins, and really doing anything I was called upon to do for the business or the family.

Holidays I spent on my Taid's boat, the *Pilot*, included trips between Mostyn or Point of Ayr and Liverpool or Widnes and between Liverpool and Rivals Head to a stone quarry, near Nefyn, in Caernarfon Bay. The cargoes of this forty-year-old, 103-ton wooden sailing craft were iron ore, coal, and stone chippings. Many a thing happens at sea, and memories of being thrown overboard, catching forty mackerel, fearing imminent shipwreck off the North Stack at Holyhead, blustery gales around the Anglesey coast, with me secured in a large coil of rope with the waves overtopping the deck, and sculling in the calm, sunny waters of the Holyhead harbour are memories that will live forever in my head. On one occasion, the overflow of stone chippings down the quarry chute nearly sunk us at Rivals. Not a joke!

Teenage Saturday nights with Bob Fazakerly, my neighbour and friend, were for me times not to be missed by a man in his lifetime. We would walk the three miles down Shotton Lane and play billiards in Shotton or go to the pictures, where we enjoyed films like *Hell's Angels* with Jean Harlow or maybe *Beau Geste* with Ronald Colman. By the time we had walked the three miles back, it would be midnight and Dad would be waiting for me with a cup of tea, usually stale but very welcome all the same.

Learning to drive came naturally to me. I have mentioned how Jim Millington tipped me off with a few wrinkles at a very early age, and since then I had always tried my hand at manoevring vehicles in the yard or on the site at Congleton, and so the gears were sort of second nature to me when I first took out a licence in December 1934. Despite the fact that I had a licence to drive a car on the road, there was little or no opportunity to utilise that privilege, as Dad seemed to live in the car himself. So when it came to pass that one had to have a certificate of competence to drive and that all those who, like me, had taken out a licence in the eighteen months leading up to the new deadline date, I found myself in the position of having to take a test, with an "L" plate on the car to signify my learner status. Dad thought perhaps, after all, I ought to get some practice in on the road, so on the July evening before I was due to take the test in Chester, he suggested I should take run through Chester, just to make sure. "What about my experienced driver?" I asked him. Well, really I needed to have someone beside me, so I took Ron Weigh to fill the front passenger seat. Off we went through Chester and on the road to Tarvin and Tarporley, but there was a slight hiccough. I was suddenly overtaken by road hypnotism because I had never been so far on the main road before. As I was going along on a quiet stretch between high copses, my concentration wavered and I wandered onto the left bank and rebounded across to the other bank on the right side of the road. This woke me up—good thing, the next day was my driving test day!

Turning up with some temerity at the appointed spot in Foregate Street, I was met by the efficient examiner, who shook hands and promptly asked me a question from the new highway code: "Under what circumstances would you pass on the left?"

My answer was incomplete: "In a one-way street, passing a tram car, and . . . um . . . umm?"

"Well, we'll leave that and see if it will come to you. Let's get on the road."

We went off and let me tell you that the streets of Chester were busy from the early days of motoring, as there were carts, barrows, and bicycles and pedestrians vying with each other for pride of place. As I was moving up Foregate Street, my attention was taken by a right hand signal of the car in front.

"There, there, that is the third circumstance in which to overtake on the left," said my passenger, in charge of my right to exist. "Go, go. Pass on the left."

I had to demur and chance failing my test. "But he's only passing a fruit barrow," I explained. "I cannot go."

It was so. The car in front did not want to turn into the side road but merely indicated he was pulling over to pass the barrow.

A successful three-point turn, an emergency stop with the Chrysler nearly putting the examiner through the window, and we returned back to base as I pulled up where we had begun in the first place. Had I failed?

"I am going to pass you, Mr. Price," said the official to my relief. "Though you have driven badly, you handle the gears very well. Go carefully in the future."

Is there a better feeling for a motorist than a legitimate certificate? Well, I had mine, but I knew my limitations!

Incidentally, you may have gathered that I have overshot the June 1934 end of Hawarden schooldays in relation to my learning to drive. For continuity, I dealt with this experience when I was in the university, really chapter 4.

However let me now interpose something about the eight months between my taking my higher school certificate in June 1934 and my good fortune in going to the university in February 1935. Why did this happen? Now let me think. For one thing, I was only seventeen years and four months old when I left Hawarden School with my highers, and thoughts of university or technological college were out of the question in those days until one was eighteen years of age. I had been unsuccessful in physics, although it was my stronger science subject, but had had no difficulty with the chemistry and Pure and Applied maths. *C'est la vie*, I thought to myself and put the idea of further education out of my mind. I thought it was about time I began to earn my own living, what with my parents' having four other children to bring up.

By now Dad was commencing a scheme of council houses, an estate of twenty units in Connah's Quay, and I opted to join him and do the officework. To begin with, I helped to set out the work for which Herbert Garratt, Esq., A.R.I.C.S., was surveyor to the urban district council. My own work on the estate contract was instructive and physically tough, but I had limited use of my school education. The officework was done in about half an hour a day, and I was bound to fill in my time somehow and this boiled down to general labouring, acting as a driver's mate, loading bricks, shovelling behind a concrete mixer, filling in foundations with rubble, and doing anything I could to be useful. Dad paid me twenty shillings a week. I paid Ma fourteen shillings, and kept six for myself, saving three of them.

At the site, labourers were paid one shilling an hour. There was a strong boy of eighteen who I thought was pretty useful and pulling his weight. I expressed my opinion to my father, and he said, "All right, Jim, if you think he's as good as a man, give him the rate." On the Friday I upped his pay from ninepence to a shilling per hour. On the Wednesday following he came into the office for his cards. The foreman had caught him wasting time and sacked him on the spot. I was distraught. If I had not been so well meaning, the boy would probably have had another chance. At ninepence an hour he might have been open to correction, but at one shilling he was on the same footing as the rest of the men. The moral of that was: "Don't get a boy to do a man's job."

Arthur Jones was a friend of my dad's, a teacher in Flint. As it happened, he was teaching engineering maths in a night school course. I must have figured in a conversation, as he came up to see me at Hendre Villa, our home. I was easily persuaded to fill in my empty evenings attending his course. The work presented me with no problems. Gradually Arthur sounded out the extent of my academic development. He was aware that two exhibition scholarships were being offered by the county education committee. A successful applicant would be given a grant towards the fees of any university in the country—in Britain, that is. After going through the necessary procedures of the application, I was fortunate to be accepted. It appeared that in the summer higher exams I was ninth in the county and only eight Scholarships had been awarded in the County in 1934.

It was a happy boy who was finally admitted to the University of Liverpool to commence in February 1935. It seemed that one had to have pure maths and applied maths as separate subjects in English universities, whereas the central Welsh board curriculum combined both subjects into one, just pure and applied. I therefore was to spend the two terms, Easter and Summer, taking the necessary maths at intermediate level and add physics for good measure, all at the university in the 1934/5 year, so effectively, I had lost no time in my continued education. I was delighted. What a lucky boy I was!

4.   Going Up

"Don't worry about them. They're not professors." It was Neil Cecil Højgaard, a fellow student who came up to me in metal workshop the first week I went up to the University of Liverpool.

Neil was a Dane, nephew of the firm of Højgaard and Schultz in Copenhagen, but he had knocked about through various schools and now was an orphan with one parent. His father was dead. Neil was a short boy with a big heart, and we became firm friends while we pursued our courses in civil engineering leading to the degree of Bachelor of Engineering (Hons.). "One day we could become capitalists," he would say. Well, I do not know if Neil ever became one. I certainly did not, though there has been a richness about my life more than money for its own sake. I have also contributed to the fortunes of others; that I cannot deny.

In a later episode I hope to refer to mine and Neil's last meeting up in Bombay in 1945, but for the present, suffice it to say that whenever I speak of events connected with academic pursuits at the university it is more likely than not that Neil was hovering around, as we did so much together, the Nordic blood of the Dane formed a real kinship with the Welshman in an English red-brick emporium of knowledge.

To get me admission into university, Dad's friend, the teacher Arthur Jones of Flint, had worked hard, and the exhibition scholarship I was awarded by the Flintshire Education Committee to attend any university in Britain resulted in an interview with the dean of the university faculty of engineering, Prof. T. B. Abell OBE., M. Eng., M.I.N.A. a naval architect, Dad, Arthur, and myself all being screened for integrity of purpose and conviction.

The scholarship payment of forty-five pounds per annum went a long way to paying the university fees and guild subscription, which together came to fifty-one pounds, three shillings, but the balance of expenses, pocket money, books, travel, accommodation at home or in digs, sports gear, and sundries all had to be found by my father, and I did not forget that there were four more children being raised after me.

Picture the journey as I went up each day to attend my lectures at the university, leaving the house soon after seven to cycle the one and a half mile to Hawarden station and take the L.N.E.R. stopper to Seacombe on the Mersey. I would sit in the corner of a twelve-seat compartment of the workmen's train, my one and tuppence ha'penny workman's ticket tucked away in the stamp pocket of my black wallet given to me by my Uncle Ste, my bike clips stuck in the pocket of my mac on the rack with its wooden rail and open, stringlike net. Here was a young man, neatly dressed in respectable ancient clothes of one-time quality, a black coat and waistcoat, pinstripe trousers, and spats, hand-me-downs from Father. The outfit lasted me for winter during my whole stay at Liverpool, although when summer came around I was fortunate to be seen about in a smart sport coat and flannels, always kept neatly pressed by Ma, who else?

Arrived at the train terminus at Seacombe, train companions and I would disgorge and make a beeline for the ferry terminus. Picture Cecil Jenkins and Tom Price of the medical school faculty, Noel Kenyon of the District Bank, Liverpool, Maxwell Davies in Liver Insurance, and Alf Sparkes, representative for Brooke Bonds Tea, scampering across the vast bus terminus at Seacombe vying with hundreds of others transferring from the L.N.E.R. train or Wirral bus to catch the ferry across the Mersey.

For two pence the *Mersey Queen* or some such craft would transport us across by the hundreds, even thousands, for they plied every ten minutes and fifteen hundred was about the carrying capacity. The crossing only took a few minutes, and then we were at Pier Head, with its imposing background of buildings fronting the city when seen from the majesty of the Mersey, the Liver Building, with its high domes up into the sky, and its clocks and its Liver birds, so distinctive, the Cunard Building, a multistorey office building, modern, with flat top, and the Mersey Docks and Harbour Board Building, architecturally fine and so right for the area.

Taking a tram, we could go for a penny up to the university via Dale Street or Water Street, through the shopping centre thoroughfare, skirting the Adelphi Hotel at one end of Limestreet and thence up Brownlow Hill to the university faculty of engineering, well up the hill on the left. At that time, the foundations of the Roman Catholic cathedral were beginning to show on the sunken area right opposite the faculty building.

The students' union was a right incline as you looked farther up Brownlow Hill from that part of the university of immediate interest to me.

That first day in metal workshop, when Neil nipped in the bud my growing bewilderment at the established endeavors of creative types like Fittock, Dixon, Orford, Stewart, and Cohen in a field strange to me, I confess to feeling like a fish out of water. For example, at the bench next

Young Jim sees life in the 1930s. (*Top left*) In Rhyl in 1932: (*left to right*) Horace, Bruce, Mrs. Price, Thirza, Iris, and Jim. (*Top right*) In 1935: two eighteen-year-olds, Harry Catherall and the author. (*Bottom two photos*)Liverpool—Merseyside—Birkenhead.

to me Fittock had a tiny model aeroplane engine in the vice, firing away, powered by lighter fuel. It was his entire creation. Everyone was centre popping pieces of metal, then dashing over to lathes and producing practical objects from the metal with the facility of a conjuror pulling a rabbit from a hat.

My own interest in the mechanical world had been limited to such disasters as stripping the gears from the gear box of Dad's old Graham Dodge. I used the term *stripping* in the destructive sense, not in its usual connotation, when a mechanic is removing gears in order to repair the box. I had been trying to rock the vehicle out of a rut in the mud on the sewage works site at Congleton. "Snatch her, Jimmy," the driver, teaching me a thing or two, had said. "If rocking her doesn't work, snatch her." Thereupon I had engaged the clutch with such severity that there had been an unearthly crash of gears and the engine revved like hades, but the truck remained stuck in the mud. Jumping outside, Bill, the driver, had looked underneath the vehicle and ejaculated, "Oh my God, the gears are all on the ground with the bottom of the gear box. The oil's everywhere." There was a sequel, which I will not introduce at this point save to say that I had to sit in the lorry and steer it all the way to Flint, some forty-five miles away, as it was towed by a lighter truck, a nightmare of a journey.

Such a thought of my destructive mechanical experience contrasted greatly with the creative works of my new colleagues, all of whom were given to addressing one another as "Mr." In fact, at the university even the lecturers and the professors addressed each student with the handle "Mister." However, from the off, Mr. Højgaard and Mr. Price were Neil and Jim to each other.

Now, for the first two terms after going up to Liverpool—in fact, the second and third terms of the 1934/5 intermediate year—one did not really have to be a professor, merely disposed to study one's subjects conscientiously, the standard being roughly equivalent to that of the higher school certificate. For physics we went over to see Professor Rice in the science faculty, and we wandered over to the parts of the university buildings to study maths, pure and maths, applied, taught by Dr. Hulme and one of his colleagues whose name escapes me, though it has been on the tip of my tongue on and off during the decades.

Professor Rice was a darling gentleman, somewhat portly, fiftyish, with thinning fair hair and florid complexion. During lectures and laboratory experiments he had a demonstrator, as professors of science subjects were wont to have, but his favourite trick was performed by himself and it is the only one I can recall offhand. He would produce some liquid air, at oodles of degrees below zero, in a stout stone jar. Then he produced a brand new tennis ball and threw it against a wall inside the lecture hall so that it rebounded, displaying an elasticity close to unity (perfection). Then he put a gutta-percha glove on his right hand, grabbed hold of the tennis ball, dipped hand and ball into the liquid, withdrew it and smashed it against the wall, fives fashion. The ball did not rebound. It smashed into smithereens as the lecture hall erupted in a loud ritual roar and the professor into a wide grin. Then, capping the spectacle, when the "All quiet" came, he stuck up his hand, minus glove to reveal two fingers off at their first joints. "That was when I forgot the glove," said Rice. There was hilarity and applause. It happened year after year, and for the gentle old man the students were henceforward putty in his hands.

One bonus I had during the two terms was to be allowed to study chemistry as part of the course without having to present it for examination. It gave the subject a sort of hobby flavour. There are those who feel that complementary subjects not directly essential to their immediate needs are a waste of time. It really depends on one's whole view of life. There were many times when I enjoyed attending a class for the sheer joy of being there, a sort of chemistry.

Dr. H. R. Hulme, the maths lecturer, was terrific. As Højgaard and I sat in the front, left side desks of his lecture room, we simply drank in his technique for expounding the mysteries of such theoretical wisdom as Fourier Series. He was succinct, simple, his cartoon-type portrayal of complex expressions so easy to follow, provided one was intent on following his every word and followed up the drift of his leads with private study. Neil and I were doing this, and we were not to be disappointed when it came to the exam results. In fact, we were both successful with our results at the summer examinations, so that apart from anything else, our time had not been wasted academically speaking.

When we went down for the summer recess there was plenty for me to do. I was beginning to have some misgivings about doing what was for me a somewhat toffee-nosed academic routine at the university, and I maintained a link with my father's endeavours in his building and civil engineering contractor line, which was what I regarded as real work. I also had a bit of a guilt complex at not earning my own living, perhaps even using up money that might have been better employed in the business.

During the summer vacation weeks, therefore, besides reading up books as a foundation to the degree course work ahead, subjects like structures, and strength of materials, and also copying up tabular data from library books, I readily helped in the business as required, mostly going up to the Congleton Sewage works contract site with Dad in his Chrysler de Soto 19.8 horsepower or with Hughie Finch, who regularly drove between sites in his Dodge 50 hundredweight tipper. I was called upon to do certain officework at the site, measuring up and working out figures for claims being prepared by Old Robbins, Dad's agent, who taught me how to use the duo-decimal system, among other things. Attendance at an arbitration hearing proved an invaluable insight into the settling up of a difference of opinion on the valuation of parts of the work that had to be carried out by the contractor. About the settlement sum my father got, well, 'nough said.

Back at home, there were figures to prepare for the houses at Connah's Quay and I had to do them.

In a lighter vein, there was a day out spent by the sea at Rhyl. I remember John Evans came up to Ron Weigh, Harry Catherall, and me, with ice creams, declaring, "They have been fighting on the Abyssynian frontier." He had seen the placards referring to Mussolini's excursion into the imperialist syndrome. We wondered if we might be called up, all of us eighteen-year-olds. In my mind, after reading a Liverpool evening newspaper at my Nain's in 1931, I felt the concept of the clouds of war gathering to bring an inevitable holocaust, forecast for 1939.

However, of immediate concern that day was a visit to the "flicks," and we were soon sitting in the circle of a modern picture house with its Reginald Dixon type of organ interlude, enjoying our ice creams in an escapist mood. Was it coincidence that we sat behind four young teenage girls? I think not! Anyway, I was rather taken with my opposite number in the seat directly in front of me.

We managed to squire the young ladies around the town after the first house evening show, then returned home on the train. The next day, in an abortive attempt to further a happy relationship, I returned to Rhyl, only to be able to divest myself of a box of chocolates to the young lady, with whom I was hopelessly in love. After coming out from the front door of her aunt's boardinghouse, she pleaded the aunt had prior claim on her time and withdrew with my chocolates, leaving me with a false address in Rhos-llanerchrugog and a deflated ego.

Fortunately, as you know, I was aware of the French proverb "c'est la vie" and my broken heart was soon mended.

When in the autumn we went up to study for part 1 of the degree course, there was an influx of those boys whose preparation had been strictly to the English curriculum, with the necessary subjects in Highers to commence the B. Eng. degree part 1 studies. There were arrivals like fondly remembered old colleagues Tommy Ithell, John Hughes, Peter Hewitt, Joe Rice, and David Statham and several others, including Messrs. Hanney, Atkinson, Nicholson, and Williamson. Part 1, B. Eng., included further maths but also introduced subjects like heat engines, electrical technology, and basic mechanical and electrical stuff, which was dropped from the curriculum for part 2 of the degree for those specialising in civil engineering. The degree courses were arranged so that one could finish up with an ordinary or an honours qualification in civil, mechanical, electrical Engineering or in naval architecture, this being Dean Abell's own specialisation. In our year, only John Burke was taking naval architecture, and so he had the dean to himself.

We had certain fellow students from overseas with names like Baghdadi, Medi Danesh Haeri, Kamal, Cohen, and L.C. Nyss, and really, for me, the presence of such men helped to give the faculty a decidedly human flavour. Baghdadi was a warm thirty-five-year-old Egyptian, Haeri a Persian who thought and calculated in his country's hieroglyphics, producing answers with great rapidity, Kamal an Arab and a serious chap, these three being of the Muslim faith, while Cohen was Jewish.

When summer term finished with the June exams in 1936, I found myself in the happy position of having performed well in "Heat Engines," having learned a particularly useful book, acquired for six shillings, practically off by heart, and, notwithstanding my "civil" intentions, was awarded for a trip to West Africa on the Elder Dempster line passenger/cargo run, which plied regularly from Liverpool and still does, so far as I know. The trip was under the auspices of the Liverpool Engineering Society. In this is I was joined by Wm. (Sos) Halewood, who was specialising in mechanical engineering but whose greater and natural talents lay on the rugby field, playing for Wigan.

To liven up my diatribe of chronological events I will now follow with a slightly dramatised account of an experience of mine when aboard the *Accra*, a motor vesel of 9337 gross tonnage, 243 first- and seventy second- and third-class carrying capacity, at a cruising speed of fourteen and a half knots.

We were in the offing of the Gambia River approaching Bathurst when it happened. I call my tale "The Tower of Terror," and it is vividly recalled after forty-five years.

The Tower of Terror

It was a clear blue tropical sky, except for a tiny dark grey spot on the horizon. That spot was getting bigger, or was it just my imagination?

In 1936, aboard M/V *Accra,* as a student engineer. Just off his eight-to-twelve watch, posing on the engineering officers' deck.

At leisure with deck quoits. Second engineer is on watch in background.

I stood alone on the engineers' exercise deck about six deck levels above the water, a student engineer from Liverpool University who had won a trip to West Africa on the M/V *Accra,* belonging to the Elder Dempster Company. The ship, of 9337 gross tons, carried over three hundred souls and had been maintaining fourteen knots as it plied around the west cost beyond Dakar, and we were now outside the mouth of the river Gambia and approaching Bathurst. It was just after midday on Thursday, the third of September 1936, and my eight-to-twelve watch was finished.

Down in the engine room I had been helping the senior fourth engineer, Bill Ashburne, reassemble a bilge water pump that had been running hot. *Not as hot,* I mused, *as we two.* Bad enough just to stand in the relative cool of the starting platform at the very bottom of the engine room with its two powerful thirty-five-hundred-horsepower D/A diesel engines, where, despite the fan ducts blowing in fresh air, the thermometer read 119 degrees Fahrenheit. Worse to go about the routine of monitoring the engine's performance or to do even a small job requiring physical effort, in my case as an extra hand on holiday just reading the oil pressures and temperatures and making myself generally useful, like I was today. Yet Jim Scragg, the senior third, had been slaving away on changing valves for one or other of the three auxiliary generators since we left L'pool. In order to emerge into the alleyway that gave access to the engineers' quarters, with their officers' mess, I had had to make my usual climb out of the hothouse of an engine room, the temperature at the exit doorway being 153 degrees Fahrenheit, like an oven.

Apart from the fact that this was my first trip abroad ever, three incidents on the journey out from Liverpool had served to emphasize the fact that this holiday would be a whole new world of experience.

A fellow student, "Sos" Halewood of Wigan, had met me in Liverpool, and we had left it until eleven at night before reporting to the ship berthed at the Toxteth dock, the day before we were due to sail, taking aboard the passengers at Princes Landing Stage. The dark, uncertain city-docks atmosphere had caused me to grip tightly the heavy nut that weighed in my pocket. My father had tipped me off of the simple device—a nut on a piece of string—to defend oneself in the event of attack from the dark shadows of the night.

Incident number one: When we arrived at the ship, we found the officer of the watch had no knowledge of our arrival. We were too early. There was no cabin available for us till the morrow, so we had to spend the night in the after quarters used occasionally for African chieftains; no heat had been through yet, certainly not to the stern of the ship, and the "comfort" of the red bunk spreads, absence of blankets, and dim

The M/V *Accra* off Madeira.

night-light was as cold as charity. The middle of the night had found Sos and I running on the spot, fully clothed, trying to keep warm in this strange icebox, our eyes glued to the door of the dormitory-type accommodation like frightened gazelles, as the silence was unbelievably strange.

Blue Peter Day, the twenty-sixth of August, and we had both cheered up by the time we had crossed the Liverpool bar, a sandbank under the narrow mouth of the Mersey. By the time we were in the Bay of Biscay, some three days after sailing, my spirits were higher until incident number two, as the slow, corkscrewing movement of the big, buoyant passenger/cargo ship synchronised with my type of anatomy and a slightly headachy feeling worsened till finally the surfeit of ship's food pushed into the inner man ejaculated like a cannonball over the side into the ocean swell, as I felt green about the gills.

Incident number three had been traumatic. As I was not at all sure of what I was to expect on the trip, my mother treated me to a white serge suit from Burton's. £3:5; 0d was a good price to pay, and I was so well aware that it was at some sacrifice that she had made the purchase. Another cheaper white cloth material at 19/11d from Lewis's was also in my gear to alternate as appropriate when I went ashore. Lying at Funchal Island, we were to go ashore for three hours at Madeira as I came out of the alleyway in my best white brand new serge. For some reason a coal barge was tied up alongside our ship; I know that because I was looking toward the port as I looked over it just before that fat greasy hawser dropped down right across my path and indelibly branded a two-inch-thick greasy black line diagonally across from left shoulder to right turnup, ruining the suit and, of course, leaving the boy inside it quite distraught.

After these searing impressions of life on this sea trip, I had been left on the qui vive. My eyes had swept the skyline when I came off duty to the relative midday cool of only blood heat as we dropped to half-speed approaching the Gold Coast port of Bathurst, and the tiny dark grey spot ten miles away on the horizon, seen from fifty feet above the plimsoll line, did not escape my sharp eyes.

The senior fourth came out of the alleyway with his white overalls and white covered engineer's cap with its black peak, attire similar to my own, and we fell into step pacing the wide deck back and to from scupper to scupper on either side of the ship.

By 'n' by I'd become certain in my own mind that the object or cloud or whatever it was, was approaching, and I taxed my experienced walking West Coaster for an explanation of what it could possible be, such a small thing in an otherwise clear sky roasted by a brilliant sun high in the heavens.

"The thing has also moved to the left of its former position," I explained to Bill.

He was somewhat dismissive of its significance. "Could be one of many things," he stated his uncertainty, which left me as ignorant as ever.

The ship was slowing to quarter-speed. In the offing of Bathurst waters she dropped anchor and swung by the bow. The port had no berthage for a vessel of our size. The breakers on the shore about half a mile away showed as a thin bolster of white foam at the front of a wash of dark blue sea below it and light blue sky above it. Some white buildings on the shore indicated the coastal town and to the right, just in the foreground, was a plantation of palms and intermingling of village hut-ments, fishing boats, and nets on the yellow sands.

From this village community an armada of tiny craft was making its untidyway towards the stationary ship. The scene, the activities, oc-cupied our attention for perhaps ten or fifteen minutes.

Coming towards the starboard side of the ship as it lay steady and parallel to the shore, the canoes had wooden outrigger stabilising tim-bers, making them appear rather like catamarans without sails. "Send the money," small boys began to call up to the passengers, now peering out from the rails of every promenade below us. From where we looked down, a large hatchway was being opened up at the main cargo deck ready for offloading, by shipboard cranes, provisions, and luggage into bigger shore craft propelled by numbers of strong oarsmen from the shore, not entirely unlike Roman galleons of old but without the slavery bit. These were free, purposeful native people.

Ashburne drew my attention to the number of toes missing from the bare feet of these adult boatmen. "They double as stevedores in some situations," he explained. "Those with more toes off have been at it the longest, shows experience." Meanwhile the little boys were diving for pennies. Sometimes they must have dived twenty feet down in the clear water in order to retrieve coins in danger of getting away.

One character was an older man. He could have been taken for the brother of Prince Monolulu. His attire, however, did not seem so extravagant, though, to the lady passengers on the ship—more outrageous, one might suppose. Perhaps not, though, as most of the passengers were either natives or dedicated West Coasters. All he wore was a top hat. He was well known to the coasters as Mr. Bones. Mr. Bones was only prepared to dive for a silver coin, usually a sixpence. It was his field to pick up the more valuable coins. When he dived, it was quite a deliberate act. He would leave his hat in the canoe while he made a five-point landing to power underwater like a human crocodile.

Having had our fill of the scene to starboard, we decided to finish our exercise before the usual taking of a shower and luncheon at one o'clock. As we turned and stepped out towards the offshore rail on the port side, the nature of the object that had first attracted my curiosity grew a little more specific. It had now grown to look like a diabolo toy turned on end. It was still in the distance but was clearly supporting a tiny black cloud. It had moved still farther to the left, but was definitely nearer and was heading for the shore in our general direction.

"Ah," said Bill, who now needed no prompting to speak. "Looks like a tornado's on its way, or . . . it could be a . . . " His voice trailed away as we both stood against the port rail and stared at the thing. "Yes, a water spout." It was like a live thing moving inexorably towards the shore but careening in a drunken, erratic manner.

Mennally, the first mate, on the first class deck below us was heard to bellow, "Hang on, everyone." Then it was sheer hell. We two gripped, then clung to, the ship's wooden rail, our legs threaded through the rails of the lower part of the bulwark, all in one desperate moment of self-preservation. A hot wind was trying to pluck us through the rails while the whole ship heaved drunkenly, its stern pointing to the awesome phenomenon in front of us, the ship like a wild horse trying frantically to be freed of its tethered head. The surface of the sea down below on the port side was seen to slip and slide, like sheets of glass, towards the spectacle only two hundred yards off. It had now taken the form of an enormous boil of water as the sea rose to a height of twenty-five feet above the surface all around, making a heavy base of water of one hundred yards diameter welling up from surrounding glassy ocean surface racing in towards it from all directions. A heavy black cloud was pouring its contents from high in the sky onto the base, and the whole dread effect was that of the double hyperbolic shape of a gigantic cooling tower built of real, solid water, hundreds of feet high.

The great prow of the *Accra* jolted to port as the omnipotent, overwhelming weight of whirling water sped onto the beach and to the native village. Boats, nets, huts, palms, and bodies were all uprooted and sucked up thirty feet off the ground into the cataclysm as it broke and finally smashed its mountain of water base over the palm forest ashore, flattening a vast clearing as far as one could see. Then all went quiet and the sea subsided with ominous calm.

Mercifully, most of the small boats had clung to the ladders and lines along the ship's side. But they were soon paddling frantically to the shore to see what had become of their kith and kin. On the ship we all breathed a prayer for our own salvation.

It was the most awesome experience I would have for a long time to come . . . or so I thought. . . .

There were sundry other events not so spectacular but all valuable experience, such as when we docked at Apapa Wharf in Lagos harbour; calling on Padre Wright; swimming on a small island in the harbour; visiting the open market; the night life, such as it was, at the Metropole, an early day/night haunt of the maritime folk; meeting an old, old man as he warmed his frail bones at the sandalwood-burning furnace oven by the premises of the Metropole, where bread, light as balloons, was being baked on the hot embers; and sailing up the Bonny River forty miles inland to Port Harcourt, which I was destined to revisit twenty years later—in chapter 13.

The five and a half weeks round trip in the Elder Dempster boat M/V *Accra* was one of the most memorable and instructive periods of my entire life, and I wish to record my respectful pleasures in looking back on it.

Not uncommonly, the ship had a Welsh captain and a Scottish chief engineer, by the name of Davidson, if I remember aright, and about eleven engineers to support the chief in his responsibilities for the proper performance of the engines.

The university organised other educational visits during term time, over and above the major practical experiences available to enthusiasts in the long vacation, and geology, surveying, and other camps in the shorter holidays.

In this bracket came nostalgically remembered places like a Coventry motorworks, the Clarence Dock Power Station, the Wigan Took Company, and the construction of the Formby-Litherland bypass, a dual carriageway, up-to-date in its time. For that construction, certainly, the resident engineer was anxious that the road would have a sweet, flowing line using transitional curves based on the spiral, for which a previous resident road engineer, H. Criswell, had developed a book of tables after supervising the great highway the Great North Road, known as the A1.

It is apposite that I mention the names of the sages responsible for such knowledge and sense that I gained when in the university, many of whom will, by the law of averages, have already passed on.

The deans of the faculty were, Prof. T. B. Abell, followed by Prof. Reg. Batson, Professor Schules in mechanical engineering, who gave us advice in the first year of the degree course proper: "Remember, you will have something to sell when you leave here." Then there was Professor Whitaker in structures, Professor Marchant in electrical technology, Professor Rowe in geology, Drs. Mawson in strengths of materials and Hulme in maths, Lecturers J. R. Daymond in Drainage and municipal engineering, Hitchens in surveying, J. J. Clark in geometric drawing, and Double in Geology. "Never trouble Double till Double troubles you" went the old adage, somewhat inaptly, old Double being a most enthusiastic and helpful asset to the geology annex in a street off Brownlow Hill.

If I were to pick on one of the sages whose lectures were of most general application to my after life, it would be Prof. Reg. Batson, who prior to his appointment as a dean at Liverpool had been an important figure in the Ministry of Transportation. His subjects of roads and railways were useful to me for about two decades of my career following university. When that professor returned after holiday recess to admit that all our notes we had taken in his lectures had been lost, he was obliged to let us all have a typed set. These were better than our handwritten ones submitted for scrutiny by Batson.

The last I saw of the dean was after his retirement several years later, by which time I was married and my wife and I met Prof. Batson admiring the sea view at Weston Super Mare, near his retirement home, in Somerset.

How about rag days? Students in Liverpool took pride in their effort each year to subscribe ever increasing sums of money to the hospitals. A great occasion was rag day, with much effort expended by all the students. Take a typical year that I recall. The celebration was in three stages. The first stage was the production of the rag mag, the *Sphinx*, with spicy entries from the various faculties, the medicals being more adept at near-the-knuckle entries. The next stage, to raise the adrenaline for the selling of the mag and the collections on rag day itself, was when we, the engineers, fought the medicals and science faculties in the quadrangle, which held our respective buildings together and apart. We used missiles in that fearsome fight. Our missiles consisted of loads of huge fish from the docks, dumped in the quadrangle. Both sides converted the dead fish into missiles of flying fish, flying backwards and forwards from bases near each faculty building until, in the end, there was a stinking mass of blooded white fish and skeletons all over the quadrangle. Since then I have seen nothing quite like it, not even when the lightwells of the Soona Mahal in Bombay were piled feet thick with garbage between times of removal by the refuse vehicles.

Rag day itself was more to the benefit of society. We all donned fancy dress and carried our collection boxes, duly authenticated with red crosses and signatures. At the end of the day, with our boxes full to overflowing from the kindly Liverpool people, we turned in our boxes and made our way home, which was not too clever, on the day of my recall, dressed as I was as a chimney sweep. Once away from Merseyside, my evening trip home to North Wales was fraught with having to withstand the black looks of the injured passenger public, who were not amused by my informal attire.

The summer of 1937 saw me doing a temporary job for eleven weeks with the county borough of Wallasey. The borough engineer was Lionel St. George Wilkinson, and H. H. Exelby was the resident engineer for the new sea wall then being built from Red Noses to Harrison Drive. My bit of engineering experience was to do with record drawings for the wall, preparation of a new drawing of a piece of realignment of Harrison Drive, and some instrument work on the beach. During my vacation, the driving of the closing pile for the cofferdam took place. Mr. Exelby, an amateur comedian of some repute, had his photo taken with the heavy steel Larssen pile dangling like a guillotine just above his head. The photo was quite amusing if not striking, but cranes have been known to slip their clutch and it would not have been so amusing to see a head rolling in the sand. For me the experience was rewarding. Out of my weekly one-pound out-of-pocket expenses I was able to afford a week's holiday with my village pal, Ron Weigh, in Douglas, Isle of Man.

We sailed out of the Mersey—correction: Fleetwood—and had as choppy a crossing as you could wish for if you wanted to see the contents of stomachs all over the deck, or the rear ends of passengers who were gazing intently down at the green sea swirling around the ship. In Douglas we two nineteen-year-olds stayed at a boardinghouse, sharing a room with other youths. On a lower floor were a bevy of beauties from Lancashire in similar accommodation. We chummed up with two young ladies, a Miss Bates and a Miss Farrington, and the transient holiday friendship helped make the well-earned break complete.

The degree year 1937/8 followed the vacation, and for the autumn and Easter terms Neil and I went into digs in the Sefton Park area in order to concentrate on studies. We then found that with travelling time negligible and free time on our hands we were able to follow pastimes, roller skating, and a course of ballroom dancing with an old dear we dubbed the chicken, though heaven knows what she called us as we trod on her feet during our excursions into the veleta and the tango.

In the final term we returned to our respective homes, probably as a result of homesickness, and I reverted to doing my evening work in the dead of night when all six of the others were asleep, but not all the time, in case of Ma. Mothers being what they are, and Ma in particular, a silent figure would invariably slip into the front sitting room as I sat at our round mahogany dining table churning out stuff for the following day. "Aren't you going to bed, lad?" was the concerned question as she gave me a hot cup of cocoa or tea to get on with. I must say the ramified daily return journey and broken sleep were a bit wearing, though inevitable.

Just a few thumbnail pictures that are flashing in about other goings on at the time I was up at the university: As we had Wednesday and Saturday afternoons off, the odd Wednesday after lectures would see us sitting through a marathon picture show at the Paramount or the Gaumont. We saw such epics as *Gone with the Wind* with Clark Gable and Vivien Leigh and *The Thirty-Nine Steps* with Robert Donat and Madeleine Carol. The big pictures, themselves nearly three hours long, would be followed by supporting films and newsreels, so by the time we had seen the whole show about four and a half hours might well have elapsed, and we came away with sore bottoms.

Lunches usually alternated between the students' union and Jim's Pub, though coffee during midmorning breaks between lectures was always at the union, where we were favoured by a bovine young waitress who would slip us an extra coffee now and again. Jim put on an excellent meal for about one and tuppence, which included a good hot main course, such as roast beef and Yorkshire pud, potatoes, and two vegetables, followed by hot apple pie and real custard and washed down with a cup of coffee—beautiful. . . . As we sat in the pub we would see a secretarial-type young lady, with personable, likeable features being treated to lunch by an older office-type gent. Two other regulars who sat beside us were middle-aged men, possibly commercial travellers, with typical scouse frankness. Neil and I were stuffing ourselves with the hot sweet course burning the skin off our tongues when one passed his open

Portrait of the newly graduated James Price, Bachelor of English degree (Hons.), 1938.

"pop-up" cigarette case to the other with the Hobson's choice of a cigarette duly proffered. The reaction of his fair pal made us explode. "Some bugger's been smoking this one," was the belligerent remark to his quite obviously responsible benefactor pal. No finesse, but he made his point in a manner that touched a chord in the students alongside who were eavesdropping (unsuspectingly as far as they themselves were concerned).

The Picton Hall Library played a part in our movements, as we borrowed books or watched international table tennis with celebrated players like Barna, the Czechoslovakian professional.

In the soccer season on Saturday afternoons I played for the third team of the university. At around five foot, eight inches and eleven stones, I played in the centre half position. We used to play local tech. or school teams or other amateurs from districts around the city like Sefton Park or Bootle.

Liverpool Engineering Society, 1937–38.

Students also spent time camping when doing fieldwork such as surveying at Llangollen. At Dolgelly, a valley was investigated as a possible site for a dam, duly designed as a practical and theoretical endeavour. Ingleton was a good place to pursue the study of geology in the field. The camaraderie of the groups enjoying all these organised pursuits was a plus added to the value of these opportunities during our young formative days.

When with examination results out in our favour we attended the induction ceremony at St. George's Hall in Liverpool, there were only two visitor tickets per student. Ma and Uncle Ste accompanied me to the hall to witness the events of this, for me, never to be forgotten day. It was typical of my father to forfeit his seat to Uncle Ste so that he might see a side of society not known to him in any particular. Ste had been the backbone of Dad's business over the years. So now the problem of getting a job was with me, as it was with every new graduate. Remember Professor Schules had told us on our first arrival that when finished at the university we would have "something to sell." Numerous applications I made resulted in three interviews, which I will summarise.

A junior engineer post was available at Keighley in Yorkshire, where a new reservoir was projected and where some design work had to be done. I was interviewed by Mr. Noel Wood, later a president of the Institution of Water Engineers.

As the salary offered was in the range of £130–70 and I really needed £200 per annum to live in digs away from home and send a few pounds home towards repayment of my parents' kindness, I declined the job.

The second interview related to a vacancy for an engineering assistant in a division of the city of Manchester drainage offices. The position was not to be offered to me until I had already accepted and taken up another job.

That job came as a result of an interview at the Royal Ordnance Factory at Chorley, Lancs., being constructed for H. M. Office of Works, the government department I was destined to join on the preparation for war effort being taken in hand to counter those smoke signals I had long forgotten about during my varsity lifetime. I left it to them whether they sent me to Glascoed, Bridgend, or Pembrey, Carmarthen, and it was to Pembrey R.O.F. that I finally started at a salary of £4 pounds a week or, rather £208 per annum.

Hence the university led me to obtaining my first job paid by someone other than Dad, and from now on, at the age of twenty-one and a half years, I would be off his hands. It must have been a great relief to my father. It certainly made me feel good. . . .

5. The First of Many

It was Saturday morning, the twentieth of August 1938, when Dad and Ma accompanied me to Chester General Railway Station in our six-year-old Chrysler de Soto 19.8 horsepower, a dark blue saloon car, with its spare wheel at the back and my leather week-end case, a twenty-first birthday present from my parents, safely guarded by Ma inside the back as I sat in the front passenger seat, with Dad at the wheel.

I had to catch the 10:18 train for Carmarthen, and as we went inexorably forward, I remember there was not a lot of conversation among us. My parents seemed to be as pensive as I was. The events of my life were turning over in my mind. My ma and dad were, beyond dispute, the two most important people in my life. They were the people who had been entirely responsible for my upbringing and my education through primary and secondary schools at Ewloe Green and Hawarden, followed by the University of Liverpool, where in June I had been successful in taking my honours degree in civil engineering.

Now I had to take up my first job as an Engineering assistant with H. M. Office of Works, a government department that was building a new royal ordnance Factory in a place called Pembrey in South Wales. Understandably the three of us had mixed feelings. We were happy, because at last I had my chance to start a job, and sad, because inside we all knew that in the future there would never again be the same relationship, that of a son dependent on his parents. The young bird was "flying the nest." In my pocket I had eleven pounds I had drawn from my savings bank, leaving twenty-five pounds balance. Never again did I expect to ask my parents for financial support. On the contrary, I felt I had incurred a financial debt that would be hard to repay. In simple logic, I assessed this at over three years of my starting pay of two hundred pounds a year. Lord only knew when if ever, that could be repaid. These thoughts were private.

Arriving at the station car park, walking over to the booking office for a ticket costing some nineteen shillings, and having a cup of tea in the buffet room before walking to the steps of the bridge over the lines to reach my platform was not an occasion for much jollification.

At last the train was ready to leave. "Look after yourself lad," my folks duetted as the guard busied himself closing doors. At the last moment on the platform, Dad handed me the remainder of his twenty-packet of cigarettes, in which there were only three Capstans left. "Here's something to smoke on the journey," he said. "Another thing, if you feel a bit down anytime, a drop of stout is often a help, in small quantities."

I was taken aback to hear Dad speak thus. Up to that time he had not encouraged me to smoke or drink. The three cigarettes were like £3 million to me, on a par with the magic of Tommy's magnet in chapter 1. At that moment, they were too precious to smoke. As the whistle blew, I kissed Ma and shook hands with Dad, holding his right hand and Ma's left.

"Let's know how you settle in, son," they shouted as the train moved away, leaving them to diminish in the distance until their waving white handkerchieves were like pieces of cotton.

Then I pulled up the compartment window and stowed my case in the rack as we headed for Carmarthen with an E.T.A. about four-thirty.

I retain but fleeting memories of the train journey between Chester and Carmarthen. To begin with, there was Edgar Cooksey sitting in my compartment, a young man I knew who lived on the A55 near the Boar's Head, not a mile from us. We were able to chat till he got out. The train was a stopper, calling at many stations as it trundled its way through Central Wales, via Whitchurch, Shrewsbury, and Craven Arms, by which time the conversation of the passengers was largely in Welsh. At that time, my knowledge of the Welsh language was limited to The Lord's Prayer, the national anthem, and a few other words.

In Carmarthen station, in bright sunshine, the buffet room was clean, green painted with mahogany counter. There was a choice of sandwiches, cut cakes, and Welsh cakes. "Tea or coffee?" a lilting Welsh accent put the question. The tea was a penny, the slab of cake two d. They went down well. The diesel train from Carmarthen to Pembrey and Burryport was soon transporting me along with a slightly swaying motion as I enjoyed the lighthearted chatter of the occupants of the large open carriage with its twin seats with backs reversible to face the engine if required.

Soon a boy informed me we would not be long. "Trimsaran is over on the left," he said as we breezed long a straight stretch with a ridge of mountain in the distance on our "port bow." "We'll be stopping at Llandore Halt next, and then it will be Burry Port. . . ."

At about five-forty, crossing the platform to the ticket collector at the wooden exit gate, I saw a familiar name on a wall across the street beyond: Buckley. That was the name of a place close to my own village of Ewloe Green, Hawarden, in North Wales. It seemed like an omen, a connection already, even if, as I looked again on giving up my ticket, I could see it was the name of a brewery company.

A single-decker red bus to Pembrey Square dropped me off at

Ashburnham Road. Remembering Dad's advice I inquired of a boy on a bike the whereabouts of the police station.

"You'll want P. C. Land. He lives just by there." He pointed to one of many similar grey rendered cottages flanking the long flat road that led on out of sight to where the village square must have been.

I opened a small gate, made a couple of strides, put my case down, and knocked at the door.

A kindly lady, fortyish, appeared.

"Is P.C. Land in?"

She looked me over as she answered my question. "My husband won't be long. He's due back any minute. Wait, you; I'll take a look." She stepped sprightly around my case, made the six-foot-long walk past me, opened her gate, and looked up the road towards the village. "Here, he's coming now."

The constable floated in on his three-speed bicycle. It had a chain guard, and the bracket over the rear wheel supported a trusty blue police cape secured tightly with a leather strap. The Lucas lamp was a fitting match for the large silver bell on the handlebars. P. C. Land dismounted, took off his bicycle clips, straightened up, and threw out his chest and his belted tummy, all in one movement. Then he took off his helmet, wiped the sweat off his thinning hair with the back of his hand, and spoke: "And what can I do for you, young man?"

"Well, Mr. Land, my name is Price. I wonder if you could advise where I might get digs. I have come down from North Wales today, and I am to start work on the construction of the new ordnance factory—on Monday."

"Oh, of course, the old powder works. Well, now, I can't say for sure . . . unless . . . Mrs. Williams, Cliff Cottage. . . . She has taken people sometimes. Tidy people, you understand, chapel ministers and such. . . . I think . . . I'll show you the house."

I nodded assent.

We walked along the roadside footpath, past the spot where the bus had dropped me off, Land pushing his bicycle on the outside of the path with his left hand, me on the inside carrying my case. We went on under a small old disused brick arched bridge, passing the Ashburnham Hotel across the road on our left, and continued as the road bent to the right over the railway bridge. There was a shriek as the train from London sped west to Carmarthen and we watched it rumbling on out of sight in the distance, a good opportunity to take a breather. "Just a few more yards," said my guide, philosopher, and friend. We were turning right immediately beyond the bridge as I learned, "That's the Golf Club up there." He looked up to his left, then waved his helmet in a quadrant. "And the links are that way."

Fifty yards more down a gentle slope in a paved lane lay what we'd come for. It turned out to be a double-fronted stone cottage, a traditional nineteenth-century house that one time surely resented the intrusion of the railway being laid alongside. The facade was plain, dignified. Sash windows were three up, two down. A home-made white glazed porch protruded with potted plants in view. The roof was of slate, sporting two stout chimneys, one at either end. It all lay back behind a couple of trees towering out of the neat stone-and-earth copse clad in white rock and turf, well cut.

We went from the gate to the front door via the path dividing the front lawn into two parts. The constable knocked in vain. "Out of luck, I'm sorry," he had to admit.

"Could you possibly suggest something else?" I ventured anxiously.

"Not really," he said, "but what I think is they've gone to town. Llanelly is only five miles away, you know. Back on the bus they'll be. Sure to be before dark." He thought a bit. " 'Fraid I'll have to leave you. If you'd like to wait here, I'll check up later to see if you've been lucky. If the worst comes to the worst, you can come back to see me."

I thought how much he resembled Sergeant Parry in Hawarden. "You're very kind, Mr. Land." Speak as you find. He was a good sort.

The law moved off up the slope from whence we had come, swaying a bit as he went. I sat on the copse alongside the gate. Half an hour later I felt it a bit cooler, took my light mac from my case, put it on, and sat down again. A few minutes later it began to shower, ever so slightly, so I got to my feet. The cooling air caused me to do an involuntary running on the spot. The rain soon stopped, but I was too damp to sit down again. Beginning to wonder how much longer my vigil would be, I saw at last someone coming into view at the top of the slope from the main road. As she approached the house, I saw it was a chubby-faced lady in dark attire with a shopping bag, sedately wearing a hat with a large brim.

As she came towards me I opened the gate, raised my hat, and allowed a clear width for her to walk through to her front door. She put down her shopping bag and turned to me. "Where have you come from? Is there something you want?"

When I asked her if she had accommodation, she replied, "Well, I do have a room. You are sure it was P. C. Land who sent you?"

"He came with me, Mrs. Williams," I confirmed.

"Come in then." She led me through a dark hallway into the kitchen beyond. As I entered I saw the floor was of quarry tiles. Opposite was a Sefton range. To the left end of the long room was a settie behind a solid square table with two chairs in place in front of it. "Sit you down over there," she said. "I'll soon blow up the fire."

I did as I was bid and sat on the settle.

Soon she came in with a tray with tea and cakes. Then she sat down on one of the chairs at the table and we had a chat. It turned out that the cottage, which was to play a big part in my life, had a very long history. It was clearly old, having been built at about 1800 as part of the Ashburnham estate. It had seen days when the tiny old harbour at Pembrey was a thriving base for exporting coal from the Gwendraeth Valley, days when a chandlery had been its main purpose and when seagoing captains had made calls at the establishment or stayed at the house. Mail had been distributed here when the house was a post office. There had been times when chapel ministers had been pleased to spend time in the house, and Wesleyans used to meet and pray in the larger of the two front rooms, quaint with its low ceilings, polished wood floor, and corner cupboards, the window fashioned into the two-foot-thick outside wall of the house. The style of the house included solid grates in heavily mantelled surrounds with large framed glass mirrors above them.

When she was a young girl, Sarah Williams, my landlady, now some sixty-odd years old, had been raised by an aunt at what she affectionately called the Cliff, originally called Cliff House, later Cliff Cottage. In November 1935, Sarah had lost her second husband, David, who had lived out his life as a maintenance engineer for the big engine providing the power at the Burry Port Tinplate Works. He had also been a parish councillor at Pembrey. Still living at the house were Sarah's brother, Roger Lewis, her stepson, Owen Williams, and her daughter, Mary Eluned.

The story of my short life was soon divulged to the old lady between gulps of tea and bites of cake, the shadows on the wall cast by the fire dancing in a hundred shapes and sizes. The strange atmosphere began to get to me. Perhaps it was the comparative age of my new benefactor and the antiquity of my new lodgings. Back home my own family was young. We were five children ranging from twenty-one to six years old. My dad had built his own house twelve years ago. It was such a vitally alive place with the builder's yard always busy with people. Here it was so sombre. I wondered at the virtue of being a roving civil engineer.

Suddenly I was snapped out of my reverie. The front door opened and closed, then the kitchen door opened to reveal a young woman slipping off her short black fur coat into one arm and taking off her grey felt hat. She moved quickly to the table. "I'm back, Mam," she announced happily. Her face was glowing. It seemed to light the room. She was the kind of person, well, her face had welcome written on it.

I am so glad Eluned became my wife. . . .

The next day, Sunday morning, after a satisfying breakfast served by Eluned in the ancient lounge that was also the dining room, warmed by a warm, crackling coal fire, I was standing on the front lawn and drinking in the fresh sea breeze sweetened by its passage over the sandy golf course with the plantation and the gorse when up came Sarah's son, Owen.

"Would you like to see the old powder works, sir?" he offered. The word *sir* was common with men in the area; this may have stemmed from the years of depression and fear of the works masters, chapel ministers, and head teachers. The old powder works had been a munitions factory in the First World War. As it was to be the site of the new royal ordnance factory, I jumped at the chance.

That Sunday walk out across the links, down past the Burrows Cottage, and way beyond the pinewood plantation and sand dunes to skirt right around the scattered old disused brick buildings of the former munitions factory was a useful preview of the place I was going to spend a year of my working life. An imposing two-story administration block still remained and was clearly in use. There was no work going on but, clearly the old buildings were being demolished, their walls having already suffered casualties, and the culprit, a large crane, with a ball dangling at the end of its jib, stood silent but guilty, ready to start up again on the morrow, my first day at work on a major site.

We were not of a mind to intrude on the site but skirted around the boundary where a tattered old wire fence was being replaced by a strong, high steel paled fence with prongs opened up.

Small fir trees, Christmas trees, by the million had been planted to reforest the area, and as we mosied around the vast site some three-quarters of a mile wide and half a mile deep, huge rabbits and hares scurried about in the thicketed area. They looked enormous to me and more like hares crossed with rats.

By the time we returned for lunch, taking care not to be clobbered by the flying balls of golfing enthusiasts heading for the nineteenth hole, my mind was so full of my new world that the emptiness of my departure on the yesterday was being replaced by the fullness of today. There was an atmosphere as we crossed the Pembrey links unique in all my world, as it was then, as it is now. Added to which there was the colourful beauty of the Pembrey mountain so visible as we ambled slowly back. I sat down to table to enjoy the first fruits of the new domestic routine I would know during the next year I was destined to remain making Pembrey my home. Eluned came in with the dinner, home-made and satisfying. Large slices of beef, vegetables from the garden in hot tureens with rich brown gravy, a pot of tea from a china teapot under a green, home-knit woolen tea cosy shaped on top with three points by the knitter's Welsh artistry, and rice pudding, thick and creamy, flavoured with nutmeg on top, all nicely browned and teeth watering.

In the weeks and months that followed I was made to feel like a person of many parts, being at once a gentleman, a lodger, a privileged paying guest, and a member of the family, and in an imperceptible way there developed a bond that was to last for all the lives of those who then resided at the old Cliff.

Now what about work, the object of the exercise? When I walked over the links to the site, some one and a quarter miles away from the house, on the twenty-second of August 1938, I was not able to envisage that it was to take me on a continuous journey in time as a roving civil engineer, gaining experience and being in site charge of so many projects along the way as my journey to retirement in the town of Barry would take me via works at Trecwn, Liverpool, Bombay, Singapore, Rosyth, Margam, Swansea, Llandilo, Ammanford, Port Harcourt, Northampton, Leicester, Peterborough, Burton-on-Tent, and Tilbury and the Surrey towns of Camberley, Ashford, Guildford, and Esher. Yes, I agree, it does read like a journey preordained for someone, almost as if I had to live out my life in a certain order, like being sent to make a travelogue, stopping on the way long enough to do a good job of casing each town and reporting to the public at large. The only difference is that my travelogue will reveal aspects of those towns that, taken as a whole, only I can tell. The journey was uniquely mine. Would that I could remember it all. It is about works, my work, and the people, or quite a few of them, who mattered to me and whose shared moments linger on as I look back on the long, tortuous journey of a roving civil engineer.

You have stayed with me up to this point. Let us continue the journey together after I walked through the gate of the site of the construction works for the royal ordnance factory at nine o'clock Monday morning, the twenty-second of August, 1938. . . .

One Inspector Jones, tall, bespectacled, grey-trilby-hatted, was making his play with the road foreman of the Holborn Construction Company as he stood on the kerbs being laid for the new internal access road and pointed out to me the way to the main office entrance.

Going into the main swing doors of the "characterful" two-storey block building facing away from the site boundary fencing and in towards the large, if compactly shaped, old powderworks area, I was cleared by the doorman and walked up the main stairs, turned left at the top, and entered the resident engineer's secretary's office. Miss Harries, petite and dark-haired, took me in to meet the R.E., Alfred Crawford, M.I.C.E., dark-haired, stocky, with bushy eyebrows, a no-nonsense figure in his early forties. I gathered that I was the first of three new university graduates to arrive on site, and after the good man had put my duties in perspective he handed me over to his deputy R.E., M.H. Maggs, Esq., A.M.I.C.E.

In no time at all I was out in a wooden site office, sharing the accommodation with an Inspector Rich, clerk of works, and Bill Lewis, the latter coming under the wing of the chief clerk of works responsible to Crawford for the buildings side of the project. Brown, the C.C.O.W., had about eight C.O.W.'s, and there were the same number of civil engineering inspectors, like Jones, Rich, Firbanks, and others, housed in site huts suitably shared, to supervise the quality of the work of an engineering nature—roads, sewers, reinforced concrete, and railways, although I must mention Tyrrell here, as he was an expert in his field, that is, railways, having been seconded from the Great Western Railway. He came from Newport, and how I liked to hear him speak with pride of his work with the "G.W."

At first I was concerned in the sewerage, both the domestic and the industrial. For the latter there were acid-resistant pipes and manholes, using Prodorite and Prodorkit pipes and materials and, in the manholes, special tiles. I was also involved in the road network.

MacLaren, chief surveyor, with Mainwaring, his assistant, had measured the six-hundred-foot-long baseline fifty-three times before they were confident enough to use it to set up a series of triangulation points to set out the works to the required standard of accuracy. My engineering checks of setting out lines and levels and my reports dovetailed with the duties of the inspectors supervising the contractors at foreman level.

The day soon came with Maggs and I were in the deputy's office and in came the resident engineer.

"We're bringing in the railways now," said Crawford. "I suggest Price be responsible for that." His suggestion to Maggs was an order.

I just listened and looked. The R.E. then took a layout plan and drew a thick red line curving in from sidings outside the boundary of the site to the plan position of the internal railway system intended for the new works. "There's a new one-in-seven crossing to be laid right there, as a temporary measure." He was sticking his finger at a spot. "We'll give them the tangent point." A new G.W.R. man named Tyrrell was coming in to see that they laid the track properly.

When my graduate colleagues Robinson and Jones arrived shortly after me, the former was allocated reinforced concrete construction work, the latter roads, while I retained the drainage and also the railways, a network of broad-gauged, flat-bottomed railway track six and a half miles long with fifty turnouts into the various magazines to store the manufactured material when the factory was operating, the magazines being bunded with sand mound protection.

The old works were quickly being replaced by new structures, among which huge iron-clad acid process buildings and concrete cooling towers reshaped a new silhouette on the landscape.

Ken Simmonds came in to work as a subagent for the Holborn Construction Company and took charge of the track laying. He was fair, fresh of face, as bouncy as Arthur Askey, a little under average height, and thirty-five. "I know you." He pointed a finger of recognition.

"Never seen you before in my life," I defended.

"No, but I've seen you. Playing marbles under the lamp by Ewloe Green Chapel."

Such a statement could only be true. It seems that Ken Simmonds was the brother-in-law of my former chemistry master Fred Roberts, who lived next door to the chapel in the village of Ewloe Green. It must have been at least five years previously when Ken had made a visit to see his sister, Mrs. Roberts, the Simmonds family coming from Bridgend, where Ken still lived. Here was a forging of a link with the past, and in no time Simmonds and I were acting in concert to get the railways in without delay.

Of an evening I would get out my Chambers seven-figure logs and compute the angles with which to establish intersection points of three lines directed from a theodolite set up on three survey stations in turn, which MacLaren had originally created for the very purpose.

"I have faith in Mr. Price," Simmonds would say in front of me when someone asked him if he was sure his setting out was accurate. This was a bit of softsoap to maintain my effort burning the midnight oil and pressing on in the fieldwork, he being by now in need of a setting out engineer and hard put to find time to stay with me as work went on apace, making other calls on his time. He was crying out for a young man to do his own lines and levels, saying to the world, "I'm a young man and I want to do well." Eventually Mr. Barber, Holborn's agent, was able to fix him up.

"Where's he coming from?" asked Simmonds.

"The Isle of Skye," was Barber's reply.

"My God, whose going to be the bloody interpreter?" asked Ken, his euphoria in having help quickly evaporating.

"But he's from Llantwit Major," Barber hastened to add.

"Oh, well, that's better."

Simmonds lit a cigarette, and blew a couple of rings, more at peace with himself at last. Even so, when David Holt did turn up on site, it was obvious that the years spent in the Celtic isle aforesaid had actually inculcated a near perfect Scottish brogue, hard to be followed by the subagent in his native South Wales. Fortunately, Holt was so sound and capable, Simmonds had no cause for complaint.

Near to Christmas, Holt and I stood together with our theodolite set up on station L with bitterly cold fingers as we pinpointed satellite station L_1 five hundred feet away, with a fine-headed nail in a stout wooden peg. It was nice to know, in the warm spring that followed, that a control exercise by a new chief surveyor, Whitaker, confirmed our nail position within a quarter of an inch.

Let us now move on from work to social and domestic life in Pembrey.

At Burry Port Memorial Hall was a social club, and I became a member, paying my one penny a week, which I paid out of the half-crown pocket money I allowed myself. At the hall I would play billiards or snooker or table tennis with colleagues from work, young engineers like Jee or Jones, and in the fine weather we would also indulge in some tennis on the hard courts in the open air. One evening there was a visit by Horace Lindrum, brother of the great Walter Lindrum, and we all stood round in the main room as the celebrity exhibited tricks of all kinds. He threw out a challenge to all comers, with the added incentive of a new autographed billiard cue to anyone who could take a frame off him in snooker. It was a piece of cake for him, as challengers fell behind, having been given four blacks. However, there was a frame in which, having broken off, one young local hero walked away with the game and the cue, to the delight and the better enjoyment of all present but to the embarrassment of the boy himself. It was, of course, before Terry Griffiths was born. Had he been on hand to play, it is possible a few clubs would have changed hands that night.

R. Brown, the chief clerk of works, discovered that I came from North Wales. The information was passed to him by his landlady, as his digs were not far from mine, on the links at Pembrey. The bush telegraph did not work too fast, as it was a long time before we realised that we both came from the county of Flint. In fact, his niece, Dorothy Bennett, had gone through Hawarden School in the same year as I. So in the early part of December 1938, I found myself in the joyful position of gong home to see my parents, and every fortnight thereafter I could look forward to a night in my old bed, after a sterling run in a new Flying Standard ten-horse power car. I was fortunate to go home for my Christmas dinner. Mr. Brown was not returning on Christmas Day, as I had to do, but a colleague, quantity surveyor Foster, had promised to pick me up on his return from the north. Ma invited him to join us at lunch, and we tucked into a wonderful Christmas dinner with all the trimmings and went on our way full of good cheer on a bright Christmas afternoon. Sadly, as we reached Llynclys crossroads, south of Oswestry, there was a collision involving a young girl on a bicycle, and this dampened our spirits for the rest of the journey. It was late after the formality of reporting the accident and then we had trouble with the lights, so it was midnight when I walked in the door of the Cliff. Sarah and Lin were relieved to see me back, though they had almost given up hope of my

Wedding photo of James and Eluned, Whit Saturday, May 27, 1939.

arrival. As they did not have a phone, there was no way I could let them have news of what had caused my delay.

It was after I saw the relief in Lin's eyes that my courtship began to develop, and at the New Year office dance at the church hall I only had eyes for Lin, according to Miss Harries, after we had been doing justice to "The Spreading Chestnut Tree." After my girl and I had walked a few times up the mountain and once to Caswell Bay on the Gower, doing thirty-five miles on that particular day, it seemed no time before we were in the month of May. We married on Whit Saturday, the twenty-seventh of May 1939, at St. Mary's Church, Burry Port. My parents had taken a dim view at what they understandably saw as my undue haste and did not attend. It was the only regret I had. I had tried to tell Ma how I felt when I went up at Christmas. When we get older, we can understand these things. At the time, we are caught up in a mood of compulsion, when priority is accorded each to his own point of view and when it is a case of not being able to please everyone.

Our honeymoon consisted of a day out at Ilfracombe. The sea crossing of the Bristol Channel from the Mumbles Head landing stage outside Swansea town was a tonic with which to begin our married life in those stringent times. We had decided to have a deferred holiday in July, when I expected to have a week off to tour around Wales in our secondhand 1934 model blue Austin ten-horsepower saloon car WN 6417, just purchased from a man in Swansea. The car had done forty-five thousand miles by the time we bought it, and we were destined to do another five thousand miles in it ourselves until it was dispensed with in the interests of balancing my books, as we will see in the next chapter. With our minds on our impending holiday, the remaining weeks spent at Pembrey and the R.O. factory do not focus easily into vision, but that canteen of cutlery with its card of best wishes from all the site personnel is an ever present reminder of the regard that we retain for all those lovely people.

Going up to North Wales in our own car via Carmarthen, Newcastle Emlyn, Aberystwyth, Dolgelly, and Harlech was a great adventure. At Harlech we stayed at a farm cottage overlooking the Cardigan Bay and Lleyn Peninsula, and the glorious summer evening with the smell of the gorse making perfume of the gentle air from the open sea was as distinctive as I had found the air at the Pembrey golf links previously touched upon. Yet here in Harlech it was different, peculiarly suited to make magic of the evening.

We were up early the next day, continuing via Caernarvon, the Menai Strait, Bangor, Conway, and Llandudno to pass through Rhyl, and Prestatyn and travel the coast road bounding the river Dee until we got to Mostyn, when my front near-side tyre blew and the car lurched involuntarily to the left, mercifully into a gateway, with an old yard area waiting to give us sanctuary from the constant flow of traffic of the main road. Out I got in my Harris tweed plus sixes and took off the wheel. Opening the shiny metal protective guard of the spare wheel revealed the spare was as flat as the tyre that had just burst. I was obliged to take the spare wheel, somewhat bald, to the next garage for repair and set off in the direction of Greenfield, bowling the wheel along the footpath as I trotted along. After about three hundred yards, I was perspiring profusely. Then I felt my trousers slip down my legs until they became pantaloons, and I was not amused at the merriment of two girls walking at the other side of the road, thinking I was some kind of clown, I supposed. Eventually, we got going again with the tyre repaired for the sum of two shillings, the going rate in those days, and continued as far as the house where I was born twenty-two years earlier.

Taid and Nain were both home and delighted to meet Lin, and we swapped anecdotes and historical data to bring us all up to date. It was not long after this that Taid would fall ill and pass away, though it happened so quietly that I was not involved with attending his funeral, the clouds of war subtly hanging over everyone's life. I have verified, however, that it was in 1939 that Taid sold the old *Pilot* or it was sold on his behalf and that the boat did not long survive him, being cast adrift when moored by Ross Island when its cable parted and running aground at Balmangan Point, Kirkcudbrightshire, on October 30, 1940.

Anyway, when we left my grandparents that day in 1939, off we went to meet my relations on Ma's side. They lived mostly in the ancient borough of Flint. Next on to our destination at Ewloe Green. When we got to the house, Hendre Villa, we were given a greeting like the prodigal son. The delay in taking our honeymoon had been well worth it, and there were family trips here and there, taking in days out at Chester, Liverpool, and Southport.

When Dad saw the state of my tyres, he insisted on my getting a new round of tyres and tubes from Davies' Tyre House in Chester, and with his introduction I was able to have a set of four wheels reshod, so to speak, for £6:10s : 0d the lot. I felt much safer on the return journey to Pembrey, which we had to make only too soon.

When we returned it was becoming clear to me that I would have to try for a post paying more money. It so happened that a vacancy was advertised by the civil engineer in chief's Department of the Admiralty for "Temporary Architectural and Civil Engineering Assistants." The advertised job would pay five pounds, ten shillings, per week, a nominal increase of thirty bob. Before the day of the interview, I burned the midnight oil working up to finality a working design and drawing of a built-up girder and a reinforced concrete beam, preparing all supporting calculations in tidy form.

On the day of the interview, I duly presented myself at Milford Haven Dockyard offices of the officer in charge of works for the Admiralty there. He enthused greatly over my concrete beam design, talking earnestly about the lines of stress set up under load. By the time the interview was over, I knew I had the job, even though the appointment would have to be made from headquarters.

I am indebted to Mr. Leigh Hunt, whose memory I cherish and respect as he was a patriot, the reason for which I shall reveal in a later chapter if you have the dogged tenacity to stay with me at least as far as chapter 9, when we will be rehabilitating Singapore. Meantime, we are off to join the Admiralty.

6.   A Change of Job

It was Saturday afternoon, July 29, 1939. I had been given my last day off, so, instead of finishing at twelve noon on this particular Saturday, we had already loaded up the car, that 1934 model Austin ten-horse-power saloon, with food and blankets and sheets, suitcases on the luggage rack sticking out behind the spare wheel case outside the rear of the car. Saucepans and a kettle were on view through the windows, and we had set off at twelve-thirty and soon arrived at Fishguard Square. We turned left into West Street, as our destination was two rooms at an address in Goodwick. Then I pulled up at the side of the road and wound down the window. I had seen someone I recognised, and he leaned down from his position in the crown of the road and rested a hand on the open sunshine roof of the car, all dressed up in its new tyres from Davies' Tyre House, Chester.

"Welcome to Fishguard." His smile was big and beaming. "I see you've brought all your goods and chattels. I live in High Street, round that way." I had met Davis when I came to visit the site at Trecwn where a new royal naval armament depot was being built for the Admiralty. He was in charge of the drawing office. He was in his late forties, permanent Admiralty, dyed in the wool, with years of tours abroad having punctuated his long service, one of the pillars of the British establishment who have given continuity and stability for centuries of history.

"We'll see you on Monday," he said on leaving us to finish our journey. We travelled down West Street and on to Goodwick to take up the "Get us in" rooms I had reserved when I came to reconnoitre the area and see the site at Trecwn. Our place in Goodwick was not far from the parish church, with a good view of the harbour, but the car had to be parked in one of a battery of sheds that served as let-out garages half a mile from the house and in the low-lying land back towards Fishguard.

In the weeks to follow we were to become fairly regular attendants at the Goodwick parish church of Saint Peter, the one social contact we made on a regular basis until we established a firm friendship there with my colleague Willie Gartshore and his family. Week-ends saw us beetling off to Pembrey as often as possible to spend our Saturday night at the Cliff. Like all regular trips of the kind, we had favourite spots where we would stop for a cup of tea from a Thermos flask. These trips were essential, as Lin was finding it difficult to relate to a new life not filled by her familiar village activities after a lifetime in Pembrey.

My early days travelling daily by car to the Trecwn Valley soon settled into a routine of sharing with three others, Davis, Gartshore, and Brodie, whom I have mentioned in order of seniority of age and standing with the Admiralty, though we were all site D.O. types—"Architectural and Civil Engineering Assistants." I remember the first day I arrived, when Joe Lumley, superintending civil engineer, received me in his office on the first floor of the old, imposing country house known as Trecwn House, situated at the end of an approach lane half a mile long. J. P. Lumley was a senior S.C.E. in the service. Closely resembling George Arliss, the famous actor, Joe deliberately packed his pipe with his stained, unmanicured Irish fingers as he meeted out his cool orders, controlling a vast site project with over sixty tunnels and magazines, mostly constructed hundreds of feet underground in the sides of the valley. The three-and-a-half-mile-long site needed a huge quarry working of dolerite rock with crushing plant breaking down the stone into railway ballast and concrete aggregate sizes. Narrow thirty-inch-gauge railway was to give access for the explosives and armament buildings that would cover the valley floor. A £4 million job in 1939 was a big job indeed, and Joe was on a "high" as he handled it deftly, a square peg in a square hole. His stock was high. Later we will meet him in different circumstances. He told me the setup in outline. Two civil engineers assisted him on the tunnels, Burnside, known as C.E.1, and Hammond, C.E.2. I was to be part of the Burnside section, with its thirty-one tunnels already in progress. Joe handed me on to Burnside. T. A. (Tammy) Burnside was the senior man of the site after Joe, a case of an Irishman and a Scotsman running the changing scene in a Welsh valley. Burnside sent me down to his site offices in a wooden building where James Wilson, A.C.E., his assistant, spent his time together with lower echelons, engineers, foremen of works, and leading men.

Wilson was another Scot, with a rasping tongue that came of wagging it too much over a very long period of time. He was a product of the Indian railways and the British Raj in India and had not descended to earth yet, since being recruited into the service at home. He suffered younger men badly, especially "immediate superior" types (whom he did not consider *his* superiors). He would try to impress us all with his sagacity, adopting a grandoise style when on the phone. "Is that Burrrrnside? You had better tell me what to do about so and so," or, "I've told them to do this," or, again, "I've told them to do that." In truth, he leaned heavily on his younger, more responsible compatriot and his main preoccupation was to keep himself in the right with the C.E.1, a good Civil Service trait.

Willie Gartshore and James Brodie were hard at it on the site office drawings, Gartshore on tunnels, Brodie on surface works in the valley, drainage, and semiunderground structures, two more Scots—the place was crowded with these dour, serious people—seeming so right for the heavy job at hand.

I was allocated a board and asked by Brodie, "Can you draft?" Admiralty type monthly progress records, inked in and duly assigned the colour of the month, had to be prepared as the work went along, to record the work and enable agreements with the contractors' engineers, like Bobbie Barnett, Thorpe, and Baker, who worked on our section setting out the tunnels under Jimmy MacBride, another Scot, a subagent for the contractors, the consortium of Pauling and Nuttal. Unlike other government departments, the Admiralty did not have clerks of works, inspectors of works, or chief inspectors to supervise the day-to-day operations in the field. That supervision was taken care of by foremen of works, and leading men, and at Trecwn there was attached a traditional clerical assistant from way back in antiquity. He was a "writer," whose job it was to commit to paper in an intelligible form the scraps of paper, the scribbled verbiage of the foremen and the leading men, so in the result a proper set of diaries and reports was available. The practice was a throwback from the days when leading men could not write, but at Trecwn, with the preponderance of work in the wet and dirty tunnel environment, there was much to be said for ensuring that the office paperwork did not carry the scars of too much realism.

Two or three days after my arrival, in came Duggan, F.W., and reported to me that he had a problem in one of the sixteen tunnels with which I was concerned. On visiting the place it was apparent that a bay for floor concrete was proceeding in the magazine chamber at the end of the tunnel. It was about thirty-six feet wide by three hundred feet long. Duggan pointed out the trouble. Concrete of a very runny consistency was oozing into the subdrain channels cut into the ground and covered with sheet iron plates supposed to act as protection to keep the channels clear during the concreting operation. I stopped the job, as I would have done on the previous site had there been a similar situation. All hell broke loose. Men came running from far and wide, including C.E.1 complete with thigh boots. Now it seems that the S.C.E., Joe Lumley, had passed an edict that we on his staff should not stop the job. We should merely condemn the work affected by the misdemeanour. On reflection, this, I agree, made good contractual sense but tended to be slower in working through the system and was fraught with the risk of action failing by default, it being inevitable that the fault would be out of sight with a battle royal likely. My action produced some kind of result. T. A. Burnside was faced with the problem of getting the job going again before wily contractor types began shouting the odds about the work being delayed and threatening with a claim. By this time a load of concrete had been dumped at an adjacent site that was not well mixed, so Tammy let fly at the ganger, telling him to demonstrate the work of spreading the concrete over the sheets spanning the drain channels below. The mix as the dollup of stuff covered the sheet left much to be desired, so C.E.1 invoked the specification to turn the stuff over twice by hand when it was in place. Jumping in with a birdsmouth spade taken from a member of the gang, he pushed in the tool the full depth of the blade, but could not effect the turning-over operation because the effort was too much for mere man. He then withdrew it a few inches. Still impossible. It was not until the blade was only submerged about three inches that it was possible to turn over the spade, making the demon-stration a flop, so now there was another problem—coarse, porous con-crete, Eventually Burnside resolved the chaos of the occasion by stopping the job himself, but I did not envy him the job of diplomacy he faced. When I returned to the office it was politely explained to me by col-leagues more knowledgeable that we D.O. types should leave the con-demning of the work to the foremen of works (who had reported the complaint to me in the first place).

If there was one lesson I learned in my subsequent decades of experience, it was that different organisations attach a different stress on the various aspects of the work required in producing their end product. In the Admiralty the D.O. types who produced the designs were the lowliest of grades in the engineering structure. In some con-sultants' offices such types are paramount. Similarly, with contractors, there is emphasis of the importance of engineers in some companies while other companies have expanded from a genesis with a practical man base, others again being developed with accountancy as the main drive. I do not think it matters so much so long as the system is sound for the organisation concerned. All I would advise is, Stick to the system. Do not think your own kicking against the pricks will result in the good of your paymasters. Of course, there is the subversive who will tell you different. That was not my way. I had seen the way of my parents and grandparents. They had thought of work as a way of life and a duty and a privilege.

One day during a week in which it was my turn to bring in my car, there were four of us up as usual. Davis was with me in the front, Brodie and Gartshore in the back. Coming in from Fishguard, we approached the entrance of the works from the side. The last stretch was down a hill, and at the bottom we had to turn left into a side turning, a sharp right angle, to join the lane giving access to Trecwn House half a mile away. We were all talking when I pushed down the clutch, revved the engine, and dropped into third. Then as we approached the turning I touched the foot brake, as was always necessary to lower the speed to that required for the turn, but nothing happened, the brake failing to operate, so I missed the turn and careered past it.

My three passengers suddenly stopped talking. Always the gentleman, Davis politely asked, "Why are we going this way?" while the car charged straight on, passing the turn into the access lane, passing the depot site entrance, and swinging right into the newly made access road from the Haverfordwest direction, which was conveniently placed to act as a runway for us to come to rest upon after a couple of hundred yards using the gears.

Up to that day, there had always been a check made of how long it took us to get to the office. Despite the fitting of new brake cables that day, future trips to work were inclusive of three backseat drivers, all ready to push their feet through the floorboards. Time no longer mattered.

Lin and I were on our knees in St. Peter's, the parish church, Goodwick, at eleven o'clock on Sunday, September 3, 1939, when Britain went to the aid of Poland, declaring war on Germany. We had hardly emerged from the church when there was an air-raid warning, that mournful wail instilling the fear of God into all mankind. Whether that particular wail was for practice or for real I do not know. Certainly the first time was probably the worst, worse even than all the times to follow when sticks of bombs would drop or be jettisoned as the pulsating four-engined bombers got to work on towns such as my beloved Liverpool or fled homewards on their escape routes across countryside like my fond Flintshire. The black clouds of war had burst!

It was now the beginning of a new era, and era of eternal night. It was a time of window blackouts and car headlight masks, black-painted, slitted, and louvred to sink all light into the ground before it could be caught by enemy aircraft. It was a time when gas masks for grown-ups, for children, for babies, were a compulsory part of attire. It was a time of rationing, of food coupons, of petrol coupons, X gallons for work and Y for private use, based on the premise that a particular type of car would do Z miles per gallon. My Austin 10 did rather better than the assessment, being relatively easy on petrol, but I was glad there was no rationing of lubricating oil, or I would have been sunk. It was no time for "Mr. Heavyfoot," and as the Austin was my first personal car, the practice of getting as many miles to the gallon as possible has stayed with me ever since. It was time to scurry off to Llanelly to join up in the sappers—the Royal Engineers—only to be told, on my return to Trecwn, that I was already part of the special (works) branch of the navy and what the hell did I think I was doing trying to join the army? They would put me in uniform when the service required it. It was a time for joining the Local Defence Volunteers, then the Home Guard, the A.R.P, or the fire-watching rosters, to defend the country wherever one breathed in the country. In the depot, we all practised shooting a .303 Enfield rifle in a disused tunnel that had been blocked up because of the treacherous ground. The fact that some of the Fishguard inhabitants thought we were building a reservoir was a tribute to the "hush-hush" attitude of the staff and workers on the site, though it was assumed that Gerry would be better informed than the locals. On watch in Trecwn House as an L.D.V., I took a walk round the rhododendron bushes with Milner, an ex-captain R.E. in the 1914–18 war, as he carried his officer's cane under one arm. Hearing a stir in the undergrowth, I moved round Milner silently, only to find I was about to have my guts removed. It was a sword stick, professionally poised my Milner. "Never get on my blind side again," he advised. "I might have run you through." It seemed he had only one eye and I had not noticed it. Not that I ever dreamed that my danger was greater from my companion than from the thing that stirred in the bush in the dead of night.

Soon Eluned and I had moved from the accommodation at Goodwick. While Lin went to stay with her mother in Pembrey for a while, I went into digs at an address in West Street, Fishguard. Lin would alternate between Pembrey and Fishguard, visiting me with her gas mask slung from her shoulder as she came off the train terminus in Fishguard harbour.

My digs were comfortable and Mrs. Hughes's cocoa before going to bed was some solace to a young engineer with a head full of the calculations of the day. Occasionally one of my colleagues, who was busy on the design and arrangement of the piles supporting the buildings in the valley floor, used to invite two of us up to his home in Haverfordwest for the evening. It was nice to experience those cherished evenings with Allen and his family in an atmosphere of settled serenity. The house was contemporary, the car a Vauxhall ten-horsepower in new condition, did its forty miles per gallon running like silk with the shot of Reddex he always believed in for upper cylinder lubrication. Brown added to the value of the evening with a tune on our host's piano, while I chipped in with a nondescript note or two with the rest of the company.

An opportunity came to rent a small empty bungalow owned by Dr. Owen, and Lin and I were planning life for ourselves in an independent home for a change. We bought the curtains and ordered a three-piece suite for fourteen pounds from a shop in Cardigan. However, the idea fell through after Lin had seen Mam in Pembrey and they decided that our firstborn, due about three months later, in late February 1940, ought to be born in the old home. The bungalow went to someone luckier than

I. It would be postwar and eight more years before we would have the opportunity of furnishing our own home again. Meantime Lin found Dai and Sophie Evans with two spare rooms we could rent in Lower Slade. It was a happy find. The address was not fancy, but it was spick and span inside. There was a marvelous rapport with our two benefactors, both of whom were of equable temperament. Dai was an engraver for a Fishguard stone mason, and the pair were natives of the Milford Haven area.

At Fishguard town, our contact with friends from the office widened to include family homesteads of Tregoning, who had seen Admiralty service in Hong Kong, and Leslie Jones, a Londoner from Portland Square. Then there was Symmonds, who kept a pub in Fishguard Square, he and his son working on the R.N.A.D. as foreman of works and leading man, respectively. Therefore I have good reason to remember Fishguard on the whole as a happy place to be.

However, Christmas 1939 was something of a letdown for me. I had to work, checking sections of the tunnels, right through the holiday, so Lin decided to leave me to it. Dai and Sophie were off to stay with their parents, so I was to be left alone. As I walked into one of the chambers I followed the gloom of a tunnel on my left hand being cut out to form a concrete footing some six feet wide and three feet thick. The concrete foundations were being put in about twenty feet at a time as the tunnel floor was excavated, in lengths, three feet below the tunnel bottom. A safety stop board was missing, and I went headlong into black, cold, icy water that had filled the hole, oilskin coat, sou'wester and all. Had not my chainman been with me, I would not be here to tell the tale, my upended position impossible to rectify by myself, owing to being snagged on an underwater timber. It was in a spare pair of pants belonging to another that I returned to my lonely rooms at 5:00 P.M. on Christmas Day, to eat cold pudding and muse over my situation with mixed feelings.

When I came into the office just before noon on Thursday, the twenty-third of February, a telegram was waiting for me: "Lin not well. Come at once." It was 1940. Gartshore, who had two children, was quick to reassure me, "Och, nay bother at a'. Dinna panic; it's just the way of life."

I got in my car and raced off like there was no more time to live, flying over every crossroads on my journey from the depot, passing through the towns of Haverfordwest, St. Clears, Carmarthen, and Kidwelly to arrive at the Cliff at two o'clock. The baby had just been born, and I knelt at the side of the bed as Lin held it in her arms the ugliest little creature in the world. *Like a monkey*, I thought.

"What a lovely baby," cooed the old lady. "Eight and a half pounds. What will you call him?"

That day we wrestled with names. Finally, it came out that my grandmother's name was Newton and Lin's mother, Mam, came out in favour of that and the matter was settled unanimously. I had three day's compassionate leave and returned to work on the following Tuesday.

Since Christmas 1939, Lin had stayed on in Pembrey awaiting her confinement, and Dai and Sophie had moved to Bristol, as Dai got a job as a signwriter in Filton, busily engaged making airplanes. I was obliged to take up new digs, with a Mr. and Miss Gray in Dinas, at a time when my car was off the road and I was possessed only of a bike, to save expense during the winter months. I began to put on weight, as the food was good and plentiful, if eventually monotonous. They must have killed a pig or had access to someone who had, for there were huge pork chops with apple sauce, ample vegetables, followed by hot apple pie and custard every night at dinner until the food almost came out of my ears. Lin visited me once when I stayed with Mr. and Miss Gray, and, now looking forward to her return to Pembrokeshire, we had the problem of shortage of accommodation.

It happened that one of the first to congratulate me on our firstborn was one on the staff who worked as messenger between the offices on the site. He came every day from St. David's, the smallest city in Wales and regarded as one of the most famous places in the principality. While on the face of it a messenger's job seems a fairly lowly standing for a man, he was an enigmatic character. For one thing, he claimed family connections with the people resident on Ramsey Island, of St. David's. For another, he used to keep popping up to London and would drop heavy stuff about how at the club he had suggested this or that to Churchill. Many of us thought he was just a yarn spinner, yet he carried himself with such bearing and had such an obviously cultured tongue, with impeccable manners, that I always felt he had been screened before being given a job that, however lowly, enabled him to carry more papers than anyone. He might well have been an adjutant in the First World War.

"I know of a small bungalow in St. David's," said Owens. "I may be able to get hold of it for you."

Lin and I were delighted when Owens was able to confirm that Mrs. Harries was prepared to let us her bungalow for "something over a pound." It turned out to be Wyncliffe Bungalow, not far from the cathedral, and the rent—twenty five shillings a week all in—related nicely to my salary, £5 : 8 : 5d less tax, at the time. Mrs. Harries, who lived in the big house next door, carefully recorded an inventory when we arrived. "I will be along each week to check my list," she warned, which immediately helped to make us feel wanted.

A few weeks after we went to live in St. David's, at the end of September 1940, we decided to sell the car. It was getting worse for wear after eight thousand miles' use by us, now having to do twenty-five miles each way to the depot every day. Returning from Pembrey one Sunday evening, on stopping for petrol at St. Clears, I asked Davo Williams if he would be prepared to buy the vehicle.

"How much are you asking?" asked Davo.

"Well, it cost me forty-five pounds so I could not let it go for less than twenty-five."

Davo lifted the bonnet and took one look. "Would you take twenty-three? It's got a welded engine. I'm thinking of my expense, friend. It's going to need some work on it."

There had been that slight bulge on the water jacket casing since I had the car. It caused a miniscule weep, but my main worry was the oil consumption. I agreed to Davo's figure. "I see your point," I said. "I'll bring it back after work tomorrow."

Monday afternoon saw me quick off the mark from work to clinch the deal and dispose of my liability.

Travelling on the A 40 from Whitland to St. Clears that evening, I stopped WN 6417 at a spot on the top of the hill looking down at the town, with Davo's garage at the bottom of the hill on the left. The car was breathing hard and I opened the bonnet to allow it to cool off a little and to wipe the weeping water and oozing oil off the side of the engine. The huge white cloth I had for the purpose turned black all over as I swabbed away the sticky black mess. Davo just beckoned me to turn the car into the corner of the forecourt and gave me a cheque without any fuss. It was a lighthearted man who caught the train back to Haverfordwest and then a bus home to St. David's.

For some time after that I had to take a tortuous route to work, leaving at six in the morning with five others travelling to Fishguard, seventeen miles away from St. David's, all chipping in ten bob a week to the owner, then having a lift with friend Gartshore in his singer 9 HP saloon from Fishguard to Trecwn eight miles away. In the evenings, Allen was kind enough to give me a lift to Haverfordwest, where I caught a bus at six o'clock for the seventeen-mile, sixteen-hill journey, which took another hour. One evening I slipped up. Therein lies the following tale.

We had arrived at Haverfordwest and Allen had dropped me at the bus stop for St. David's. As the older generation will remember, in those days busses did not have destinations displayed. Signposts were removed from the roadsides. It was all in aid of national security, serving to confuse any invaders as well as our own population. At about one minute to six that night, along came a bus and I made to get on it. However, as the bus did have its destination Carmarthen, displayed, I waved it on and off it went, but it did not turn right around the block as it should have done; instead it carried on towards St. David's. I went cold as I realised it *was* my bus after all, the last bus home. I had to walk. There was nothing else for it. It was going dark and in the gloomy darkness no one wanted to give a lift to the black figure dressed in sou'wester, oilskin coat, and gumboots with thick socks, who might easily be a paratrooper. Curse it! I walked on at three miles an hour. Trudge, trudge, trudge. Seventeen miles and sixteen hills to go. Black night, heavy clouds, moonless, steady rain on my face, my only refreshment since I had eaten my sandwich lunch before twelve-twenty that lunchtime. The minutes grew to quarter-hours, half-hours, hours. I began to feel like a Pilgrim walking through the valley of the shadow. . . .

After ten miles on the road, I paused and gazed from the top of a hill leading down to a seaside village, the phosphorescence of the sea just permitting me to make out the shape of the settlement half a mile away.

After pulling up my thick socks, which had slipped into my gumboots beneath the black leggings, I stood up and stretched my arms outward and upward to flex my weary muscles and breathe in the fresh air wafting up from below. The shadowy figure of a young man pushing a bicycle approached up the hill, a silhouette signifying a fellow soul on the earth. But he did not think so. As he looked up, he evidently took me for a messenger from Hades or Hitler, which most would say were synonymous, and let out a shriek, parting company with the bike as he darted to one side to avoid me. As the penny dropped, I hastened to reassure the young soldier, "You all right, chum?"

He regained his bicycle and walked quickly onwards without stopping to talk, emitting a thin whistle as he went on his way. "Good night," I spoke again, and this time he called back, "Night, sir," as he melted into the night.

The incident rejuvenated my flagging form, and from that point the remaining seven miles home were lit by the memory, my head taking over the control of my sore feet as I reflected with some amusement on what had happened. Then the journey was over. It was a quarter past eleven. Lin was beside herself and not at all amused at my lame excuse for being late or my sense of fun at the anecdote about the soldier or my chagrin at seeing two huge holes in the heels of my socks and the skin peeled off underneath. A friend had kept her company for nearly four hours once it had become clear that I was not on the bus.

In November 1940, my arrangement for getting to work fizzled out and I had the offer of a motorcycle from a colleague who lived in the country at Trevine near the A 487 coast road. The 1931 350 cc OHV Ariel was a heavy machine, about three hundred weight. At the farm where he lived, my friend took the six pounds we had agreed upon; then I sat in the saddle gave the kick start all I'd got, and it roared into life with the twist grip halfway out on the handlebars. Letting in the clutch, I zoomed up the hill towards the main coast road, but conked out halfway up and staggered back to the bottom again for my waiting friend to give the thing another kick, and taking his advice to give it plenty of juice, I was second time lucky and negotiated the top, swaying from side to side. When I arrived at the front door of the bungalow, the machine fell on me and pinned me to the ground. Our baby son, standing on a chair, called in his baby way, "Nin, Nin," which brought Lin to help me right the thing. Cumbersome though my motorcycle was, I learned to handle it better in the course of the next few days. During these days I was alerted to my problem of handling when, coming from Haverfordwest, I was obliged to retreat backwards from a bus bearing down upon me on the horseshoe bend of one of the steep hills.

Some will remember the cold winter of 1941. Black ice was a bugbear in the area we lived. Off to work one morning, I was alarmed to feel the bike continue straight on as I attempted to bear left on a slow, easy bend on a flat stretch of road. I slid to rest on my side, leggings dragging along the slippery road surface. Lifting up the bike, I went on to do the same thing as I tried to negotiate the next slow bend, this time the gear box being affected. Leaving the machine at a roadside farm at Mathry, I thumbed a lift. In the evening I was dropped off at the farm but failed to start it and waited for another lift home. Some of my colleagues turned up from the office and told me there had been a telegram at the office asking me to meet Lin off the bus from Carmarthen arriving at St. David's at seven o'clock. I was just in time to meet the bus. Lin, Newton, and fourteen packages disgorged from the single-decker bus onto the pavement. How had it been possible for her to enlist the help of so many passengers and transfer the loads from bus to bus at Haverfordwest I cannot imagine. Lin explained she had curtailed her projected week's holiday by three days because she had dreamt I had had an accident on the motorcycle.

Meanwhile, what had happened to me careerwise? In the summer of 1940, the current superintending civil engineer, F. G. (Freddie) Brighton, was looking in at the divisional office when I asked him for a scheme to prepare that I could present to the Institution of Civil Engineers to accompany my final part C examinations leading to corporate membership in the institution by the time I reached the minimum age of twenty-five.

"Here, Jamie, here, Jamie," Freddie said. It was something he said incessantly as he went round the site. He was talking to his reddish brown spaniel bitch, which ran all over the place. "I don't see why not." He was now talking to me, using his other much repeated expression. "You can do the depot road." He had been wondering who to unload that job onto, so I had the whole of the preparation of the new depot roads' scheme. This entailed three and a half miles, which had to wend its way through the buildings in the valley floor, and another half-mile to form the new access up to Trecwn House. The next six weeks saw me out in the field armed with the necessary surveying accoutrements, attaché case full of papers, Chambers seven figure logs Trig. tables, and Cooke, Troughton and Simms's theodolite, but most of all, my faithful chain-man, Tom Harries, with the tools of his trade and an invaluable sense of touch with the steel tape. The Royal Army Ordnance corps had once established a grid of pegs over the territory, only three of which remained as a means of reference, and it was a relieved young man with aspirations of becoming a professional person who saw the closing of the open traverse within 1.5 inches of latitude and 5.5 inches of longitude. The special effort made for three months that summer surveying, setting out, making calculations, drawing, taking offsheets, filling bill of quantities, and making specifications meant that I could sit the final examination for corporation membership in the following year, 1941. In the meantime, Freddie Brighton had been replaced by P. R. Robinson, who kindly endorsed my papers for submission to the institution and, at the same time, recommended me for promotion to A.C.E. (assistant civil engineer), which would lead to the next major period in my life.

Lin and I had a tremendous sendoff, with a social evening at the Fishguard Bay Hotel, all Admiralty colleagues and wives attending. It was midnight when we eventually took our leave of the party. I had all their names on a piece of paper, but I was torn by my feelings, happy at the prospect of being promoted to my beloved Liverpool as from the twenty-fifth of February 1941, but sad at the thought that I might never again see any of those Fishguard friends. I never did, except in my mind. . . .

7. Onward and Upward

On Monday, February 24, 1941, I travelled from Fishguard, South Wales, to arrive at Lime Street station, Liverpool, bleary-eyed and cold after spending three hours in the middle of the night at Crewe station, the coldest night I had known, even including my first night on the *Accra*. There had been no tea, nothing at all, on the journey, and in those war days you could not buy a Thermos flask to help along when travelling.

Having wasted no time reporting to the Admiralty in the Liver Building, I saw the departmental higher clerical officer, Jordan, who ushered me in to meet the important-looking man behind a bright new desk in a large room on the sixth floor that overlooked the mouth of the Mersey. The desk was situated in the middle of a square Axminster carpet, befitting a man of his standing as superintending civil engineer, the equivalent of Captain (Special Branch) R.N.V.R. I stood on the corticene flooring lino fitting over the whole square room. Higginbotham beckoned me to take a seat opposite him at the desk. Almost holding onto the desk, I must have cut a very poor specimen; underfed, sleepy, shivering slightly, with a cold contracted on the train journey. Heaven knows what he thought they had sent him as an A.C.E. Even two years later a little bird had told me he had referred to me as being of "extreme youth" even if "sensible."

"I'll put you with Littlepage. He's C.E. Lancs," he informed me.

Littlepage came in at the summons of his bell. "This is Price, just arrived from Trecwn. Take him under your wing," Higginbotham told him.

Both men were bald, with high domes, but Higginbotham was the taller, fairer, and younger of the two, Littlepage being a Jamaican, though his swarthy complexion did not betray his origin. His jovial face would have fitted in anywhere.

At one of the two desks in room 625 and in no time at all I had learned from Littlepage that our area covered practically the whole of Lancashire, with other naval shore properties as far afield as Newtown and Holyhead in Wales, Workington in Cumberland, and Sowerby bridge in Yorkshire. Meantime he asked me to knock out a drawing for a rail platform for a siding into the Palatine Works at Warrington, while he nipped out to see some design going on in the drawing office along the corridor on the same floor. I am moved to write of what ensued to give me a crash course in procedure, a lesson in the chain of command, in the form of a story, true in the main, nautical but nice. . . .

Chain of Command

"He's a terrible man," said the voice over the telephone.

What a way to start my duties. I had just answered the telephone in my new office in the Liver Building and been baffled and bewildered by the caller's conversation in code. A minute of mortification had led up to the Parthian shot. Fortunately, the phone went dead before my *regia mortis* set in finally, leaving my body unable to tell the tale. Newly promoted to Liverpool from Trecwn, where the Admiralty had been doing a solid job of work blasting dolerite rock to fashion a large armament depot in a three-mile-long site in a Welsh valley, I was now floundering in a sea of papers, including fleet orders, covering my desk trays and my know-how of my new task was nil. It was the naval procedure and language that I lacked. Nowadays you go on a course. Then you found out for yourself. I went to the window of my office on the sixth floor and could see some more normal people walking up the gangway of the *Apapa,* about to embark from Prince's Landing far down below. The only memory I retained of the homily over the phone was the name, Campbell.

North Western Command Headquarters occupied the whole of the Liver Building, which still proudly dominates the city of Liverpool when seen from the river Mersey. Occupying the sixth floor was the Admiralty Works Office, set up by the civil engineer in chief to cover the works required for some 150 naval shore establishments in the north-west. I was now an assistant civil engineer, an A.C.E., still in civvies but with the equivalent standing of an R.N.V.R. "wavy" navy lieutenant, Special Branch. Permanent officers were endowed with straight stripes. In my newly discovered ineptitude I was bound to feel I had no standing at all.

In came my immediate superior, the C.E. He was Harry A. Little-page, but quite a big page at that moment compared to me. "Oh, it was Commander Campbell, the old rogue," he said with a crinkly grin. "He was an admiral before he retired. You've really plunged in at the deep end. Now you'll need a life belt." He dropped onto my desk a hard-backed blue-grey book and then went out again to fight the war. The title was *Instructions to Officers in Charge of Works.* I thumbed over the pages and could see immediately it was a sort of service bible, "gospel according to the navy." My first thought was to put right my linguistic and procedural deficiency, forgetting the fact that my in tray was already beginning to build up with routine work matters. After an hour or so, Campbell's codified style of conversation was becoming clearer. There were rules to define where one man's duties started and finished, clarifying authority and responsibility for each category of person in the department, as well as short forms for titles in the naval service, both for animate and inanimate things. I left the office late on my first day of duty, full of FOI/C, NOI/C, OCW, and various abbreviations, thinking respectively of flag officer in charge, naval officer in charge, and officer in charge of works. Nine o'clock saw me descending the lift and going along the main corridor on the ground floor, walking past the sentry on duty watch at the main entrance wishing him good night and "TTFN" slip of the tongue quip. I slipped into the Water Street night.

The next day on my, so to speak, maiden trip from the Liver, there were two of us trotting out of Workington railway station facing a bitter cold day in February, the spectre of the voice on the phone, its face and hands shapeless, still consuming my subconscious mind. G.F. Marr, F.W., was my foreman of works and right hand man on the trip, and he looked the epitome of confidence—dapper, with gold-rimmed spectacles, an experienced Scotsman. Long before the arrival at our destination I had formed the view that he was just the type of man for his current calling, flitting about from site to site dazzling the dense with his official instructions from above and despatching those orders with the certainty of a salesman, which he may well have been prewar.

I was half his age, my seniority of status paper thin as I soft-pedalled alongside him, plan in hand, through the main swing doors of our rendezvous with the naval officer in charge. The NOI/C dominated the deck of the hotel foyer, standing there, massive, mature, and larger than life. Introductions over, I was relieved to realise that Campbell was quite unaware that he had already had a parley with the young person before him. *Must have thought you were a cleaner,* consoled my conscience.

"The contractor'll be here in a jiff. We'll all go down together." It was a statement of fact.

Hobson, the local builder, came in with his donkey coat spattered with dry concrete, his face blandly bountiful, his hands coloured by his dual role of a working master. "He's real enough," said my mentor. Hobson's offer of hospitality was declined by the old salt, and we wandered down to the dunes by the shore. Marr spritely stubbed his heel into the hardpan surface he sought out on the beach. Stationed at a corner apiece, we perused a copy of the plan of the concrete box proposed for the new lookout post.

"Just right," said the commander; then, chameleonlike, he changed to the guise of a Lord High Admiral. "I'll have it fifty feet up in the air." Hobson was delighted, dazed with disbelief, a fortune in profit prospected.

Marr choked in a stage whisper to me, "It's not a *ruddy light house;*" then, in unguarded affront to the commander, he stated the obvious: "The plan shows it on the ground."

"No use to me, Marr, if I can't see a ten-mile horizon!" C——exploded, with all the exasperation he could muster as a man who had sailed the seven seas.

It was a testing time for me, suffering somewhat and needing a sop from Solomon. "If the *admiral* wants it fifty feet up in the air, then fifty feet up in the air it must go." The former admiral beamed with content as I went on: "All we have to do is report the admiral's decision to H/Q. We'll ring up the M.C. on return to the hotel. We're all happy with the site."

Back in the hotel the commander was quick to command, "Ring the Liver. Let him know our course of action."

"The M.C. would take it better from you, Admiral," I said. It was I who rang the Liver.

Littlepage went down to see the maintenance captain on the first floor. He managed the flag officer's liaison with the works department.

A quarter-hour passed as we waited around in the foyer; then Campbell was paged and answered a call to the phone. He returned with an expression giving nothing away. "I'll make a start right away, Admiral," said Hobson hopefully.

"I have told them it will do me on the ground, Hobson, my dear chap, and another thing, not so much of the 'Admiral.' I'm a flipping commander in this war. Let's all splice the mainbrace." "Mine's a rum, please," I ventured.

"You're a man after my own heart, Price," he said, slapping me on the back. "We'll both have a double."

Everyone followed suit with a double rum. It was a good time to fall into line with the good old had-been.

On the way home in the train, I closed my eyes contentedly. I had had a two days' crash course in chain of command.

My crash course on procedures was to stand me in good stead in the months and years to come. Just to introduce you to the atmosphere of the Liverpool office, let me draw you a pen picture of each of my superiors when I first arrived at the place.

Roger Wordsworth Higginbotham was the superintending civil engineer or S.C.E on seat, in charge of all works for the naval North Western Command. His high dome and intellectual brow suggested that this M.A. Cambridge was a man of great mental capacity. He dispensed his duties with an air of efficiency and authority. Some seventy staff supported him in supervising works, mainly civil engineering in content, prepared in the office and executed by contractors.

Harry A. Littlepage was my first immediate superior. He was the divisional civil engineer (outstations)—a most friendly, cheery, cherubic chap, a credit to his Jamaican homeland. During the time we would work together, his other assistant being Arthur Lang of Dewsbury, Yorkshire, there would be no sense of stress in the divisional office. Occasionally we would be joined at lunch by one of his countrymen whom he called Cons. It was none other than Leary Constantine, the great West Indian test cricketer, who years later was to become Sir Leary Constantine. At the time, Cons was employed as a labour employment man in Liverpool. Littlepage would eventually return to his own country as Director of Housing, West Indies.

The job of an assistant civil engineer at the Liverpool Works Office was a busy one. In my case it meant making trips every other day round the extensive area, with so many "shore establishments" dotted around to service the navy.

There were naval store depots, victualing store depots, rum depots, various personnel accommodations, and places of special functions, W/T, radar, minesweeping gear, and others calling for individual special arrangements for all the various establishments.

Drawing Office "Architectural and Civil Engineering" people would help to survey a property and detail out each scheme and prepare specifications. Quantity surveyors would prepare estimates and any quantity of bills required. An A.C.E. would coordinate the preparation of a contract, drawing up reports that, after vetting by the C.E.-S.C.E route, would go up to headquarters for approval.

It was a system that worked. It impressed me all the more because it had the force of tradition behind it.

In all the subsequent years since leaving the Admiralty in 1947, and particularly when I left Liverpool in 1945, I have met up with various systems in public and private organisations and each have placed emphasis on different aspects of the work, but I always found that for a system to work it had to have gone beyond the "growing pains" stage and once the system was established the concern worked best when the staff had the loyalty to operate it to the best of their ability. Of course, this does not preclude improvements being made as weaknesses become apparent, provided these are not made in a vacuum to further the ends of the rebel or the rabble.

There in Liverpool I was just bedding in to my ACE job during the first year or so spent under Littlepage's wing.

Various minor works were accomplished, mostly converting old

Lancashire mill premises to naval purposes, and eventually things were settling in to a six- or seven-day-a-week slog. In June 1941, only a short time after my arrival, I sat my AMICE, part C, exam to back up my application for the corporate membership in the institution. My day-to-day work in Liverpool was of direct help to me in answering the exam questions.

In 1942, while at the Admiralty Works Office, Liverpool, my report to the University detailing my four years outside experience as a civil engineer resulted in my admission to the degree of Master of Engineering.

On the home front it was not exactly a picnic for any young married people at the outbreak of war or, for that matter, for any of the population, for we were all involved with the cataclysmic confrontation with the enemy bombardments that have been written about ad nauseam. I do not think my personal angle on the Liverpool air raids would serve to embellish the graphic accounts available in abundance.

At the outbreak of the war, private housing was brought to a standstill. My father had a scheme of twenty houses under way, and because of the lack of materials for private work he was unable to proceed beyond first-floor level with several of them, which meant, in modern jargon, "cessation of cash flow." Timber certificates, steel certificates, petrol coupons, and food coupons had to be sought from government controllers' offices, and it was much easier to stop than to go. I used to comment to some that it was like pushing over fourteen-inch-thick brick walls to get anything done. That was the feeling that most of us "temporary bastards" had in the service. Yet, in retrospect, it was perhaps because of the frustration we had that the walls were pushed over by all those millions of people strange to the war-effort situation, all those "temporary bastards" everywhere who won the war.

Now because of the impossibility of house building, accommodation for the migrants in the community became impossibly difficult, as Lin and I discovered to our constant chagrin and despair. For a couple with a child trying to find somewhere to live was like looking for a needle in a haystack. Some Sundays I would be trekking out to outlandish places following up some lead that always became a dead end, others having beaten me to it. Eventually, our quantity surveyor, Knox of Oban in Scotland, had to go off on three months' convalescent leave through overwork and understaff and we had the chance of his modern three-bedroom semi in Huyton.

Those three months in Huyton went all too quickly; then Lin returned to the safety of life at the old home at Pembrey, while I travelled up daily from Hendre Villa, my father's home in Ewloe Green, Hawarden. There were spells together in which we lived as a family in my parents' home, but these were uneasy periods, as the Gerry bombers always seemed to return over the house after they had completed their sorties over Liverpool and on one occasion dropped a stick of bombs uncomfortably close in a neighbouring field. Air-raid warnings for our family at this time meant diving out to the improvised shelter in the washhouse outside in the yard in the cold of the night, on one such occasion Lin having a miscarriage of a three-month old pregnancy.

Apart from the times spent together already mentioned, just a few weeks at a time, we did spend one full year spell together out of the four years I was on duty at Liverpool. This was when we managed to rent a furnished bungalow in Moreton, Wirral. That was a furnished property, small, full of heirlooms and relics and ornaments, the landlord being an old widower sentimentally disposed to call every week to check his inventory just like the landlady for the bungalow at St. David's.

James Price, Master of English degree recipient, 1942.

In the time I worked at the Liverpool office I served under six divisional civil engineers for outstation establishments and three superintending civil engineers in full charge. The "itinerant ones" numbered the C.E.'s Littlepage, Clark, Oxley, Knowles, Lawson, and Winters. The S.C.E.'s were Higginbotham, Lumley (formerly of Trecwn), and Lowe.

My experience under A. H. Clark was noteworthy. I was with him for a year when Littlepage took over other work for the S.C.E. Mr. Clark was a formerly retired gentleman of sixty-three years from Goodwick, Pem., who had obtained a job with the Admiralty to do his bit for the country. He had boundless energy, always bowling in and out of Higginbotham's office with schemes and specifications and estimates prepared by his assistants, Harrison, Lang, and me, and of course we had the help of the drawing office, Lands, and Quantity Surveyors' staffs. A. H. Clark would busy himself, one day in and one day out, around the establishments in Lancashire. When I accompanied him I noticed how he would often barge through the traffic lights when on red as he went along in his grey, newish Ford 10 Beetle car with its red seats.

"Oh, did I?" he would say when I mentioned the fact. "Well, it's too late now; we're on our way."

When he lost his way I once inquired how he ascertained what mileage to charge the Admiralty.

"Depends whether you think it is me or the Admiralty losing the way," he replied noncommittally. He had reversible spectacles for reading and long sight, as he had lost the sight of one eye and in the office, he would swivel the arms of the specs and turn them around suitable for distance focussing whenever he looked up at anyone across the desk. Then the process was reversed as he looked down again at his papers. He sported a small pencil with a metal cap on it, in scale with his handwriting, neat and small characters, so many to the inch of pencil length.

When this young man picked up the tiny tool of his thinking trade, there was a loud Clark cry: "You . . . kleptomaniac!" I am sure the pencil was one he had used in his school days. . . . While Clark was in charge of the division, a lot of work was pushed out of the office and supervised, but after a year, he went back into retirement, and his successors were to be quite a procession in the final two years of my time at the Liverpool office. Although I have given special mention to Littlepage and Clark, I was grateful for the guidance given by all my seniors at Liverpool and for the help of those who worked along with me, not only the technical and practical types but also the administrative and clerical ones.

The foremen of works trip lightly onto the paper with the pageant of colleagues of the time, men like Marr, Wilson (an Irishman, much given to the use of "an R.S.J." and or rolled steel joist, as the elixer to cure all problems of alteration of buildings), Welton (a seventy-three-year-old ex–general foreman for the Yorkshire Hennebique reinforced concrete company, now claiming to be sixty-three in order to do his bit for the nation), and Lord, who was quick to volunteer for service overseas, having been dazzled by letters of those who had witnessed the beautiful blue of the Mediterranean.

There were contractors galore, people like Gilbert Collier of Bolton, Thomas Croft of Preston, George Dew and Co. of Oldham, and C. Seward and Co., Heating Contractors of Preston, springing to mind. In the drawing office, stalwarts like C.M.G. Keeping (with his prewar twenty-nine-horsepower Railton in wraps for the duration) and Gibson, also big, but thin on top and in civvie times a brewery engineer, these two contrasting with their close colleague and friend C. V. Jones, diminutive, dapper, and precision personified. Halloway was a case of one promoted at Liverpool in mature years from the drawing office to join us in our division as an A.C.E., which was especially nice, as he had a big frame to match his shining pate and jovial face and, most of all, he and his wife were friends of my Auntie Doris in Flint, whose husband, John (Jacko) Birch, was an accountant who had shrewdly invested the money made from her flair in the confectionery business, so that she was left secure for life with a good deal of property to her name when John died at a very early age.

The nature of the varied works is worthy of recalling with examples illustrative of what had to be done and related personal touches to paint the pictures of our lives at the time. We'll look in on four establishments—Watersmeeting Mill at Bolton, Underwood Mill at Oswaldtwistle, the Detention Quarters at Preston and Castle Mill at Oldham.

Watersmeeting was one of two ancient mills in Lancashire converted to rum depots for the navy. It had been a bleachworks, or something of the kind, before it was requisitioned by our Lands Branch and a scheme drawn up in our drawing office by C.M.G. Keeping for its new function giving succour to the boys in the Senior Service. I remember the stills, the vats, the victualling storemen in charge of these purveyors of perfect golden, syrupy, heavily overproofed liquid and the substantial sips spared from the samples tasted by the connoisseurs responsible for the quality. Strange that a schooner of such perfect potency left the head still clear and the stomach satisfied. 'Nuff said on this delectable dream.

At Underwood Mill in Oswaldtwistle there was decay in the top rings of its chimney of dizzy height. Old Welton, my F.W. in the area, and at seventy-three nearly as old as the chimney, was not in such a decrepit condition. He always insisted on carrying my briefcase when he met me off the train at any point. "It keeps me balanced, sir," he explained. We gazed up at the light ladders dogged in to the side of the chimney all the way up to the top. The steeplejacks were just refixing the lightning conductor after finishing the new masonry rings, men experienced in the work. From the ground it certainly seemed a neat job of work. Nevertheless, Welton, with a lifetime of experience in high reinforced concrete structures at home and in South Africa, said in his Gravesend Thames accent, "Do you mind holding your case? I'll just pop up and have a look round at close quarters."

"Like hell you will," I contradicted. "I think it will last as long as the Admiralty needs it." I did not want it on my conscience that old Welton had been allowed to break his neck, too high a price to pay just to see how pretty the masonry was well over a hundred feet off the ground. Nevertheless, I had to make it a firm order before I could hold the old foreman back. Of special interest to me is the comparative high value for money that steeplejacks gave for the hazardous work they did.

At Preston, the old disused prison was site for resurrection of three of its five wings to provide a naval detention quarters to accommodate over seven hundred "delinquents"—a euphemistic term for prisoners. The Admiralty appointed an old commander, Roe, as liaison officer during the preparation of the contract and right through to the completion of the work after the contract had been let. There was a whole gamut of work, entailing installing galley, church, and boiler room with a new heating system for this old "glass house" type relic of the past, which had been formerly heated by hot air, blowing at high level into each cell, the air cooling and returning back from the floor level grills to the control centre of the cell blocks arranged around a centre block like spokes round the hub of a wheel.

Two of the five cell blocks of the original corrective establishment were excluded from the scheme by the building of a high brick wall atop of which was fixed a coil of Dannert barbed wire, which those who can bring it to mind are likely to put it in the category of the Berlin Wall built after the war.

When Lawson and Halloway were working on the estimate for Preston Prison conversion to detention quarters, it was anyone's guess what individual cells would cost to refurbish, they being in such a tatty condition. Lawson, the then C.E., would come out with thirty pounds ("Thairrty pounds," in his Scots' accent), and that amount was introduced into the estimate numerous times for the nebulous operations involved in the resuscitation of the whole assembly of run-down buildings in the first stages of decay.

In the result, however, it was surprising how near the internal estimate of approximately twelve thousand pounds matched the sum in the lowest tender accepted for the work. Commander Roe had a difficult job trying to advise on requirements in the planning stage. Imagine a surgeon trying to advise on building requirements for a hospital, and you have the analogy. The specialist for whom an intricate establishment is required always knows what he wanted when he finds he has not got it. When the prison contract work was proceeding, therefore, Roe could picture the fallacies of certain things being done and asked for changes to be made which always adds to the cost and to the time of the work. Even after the new detention quarters were opened, my F. W. Wilson, was kept busy on this particular location, where he called in every day on his rounds of that part of Lancashire with which he was concerned. I was glad that Wilson, a native of Belfast, was a diplomat. He kept his cool while the place was gradually nursed into the state suitable for the commander. There was the case of the young delinquent sailor who was given solitary confinement. This meant he was being punished for making his escape over the roofs of the prison buildings and by his own ingenuity getting far away, only to be picked up somewhere on the East Lancs Road. When he was left in the isolation cell, safely secured for the night in a straightjacket, a kindly prison officer eased the top lace of the outfit to let the boy breathe without too much discomfort. In the morning, a peep through the spyhole in the thick cell door revealed no one where someone should be. When the door was opened the young Houdini stood to attention by a wall, his straightjacket on the floor, neatly folded beside him. In the silence of his isolated cell in the night he had also been hard at it trying to make his escape via this strong basement-type cast-iron window with its tiny pavement-type stout glass panes, its cast-iron glazed bars forming a heavy, impenetrable grill. Two rows of bricks had been dislodged from the inside cell wall near the floor level, the mortar scraped out by some mystery tool, never found. Bricks had been placed to make a working platform under the small window, high up in the end of his basement cell, and he had been in the throes of smashing out the stout, chunky glass. During the time of his overnight stay he had not succeeded in smashing out the heavy cast-iron framing grillage. No doubt about it, given the long solitude periods commonly endured by prisoners like the Count of Monte Cristo on Devil's Island, our young fireball, some mother's son, would have made his escape. The boy had used a piece of his prison tunic to wrap round one glazing bar to hang on to for leverage while he smashed out the glass with the other, using bricks as hammers until they broke, in turn, the glass and the bricks.

Much as our young friend earned the admiration of everyone, he had more solitary for his pains, with the top lace tied securely, and was charged the cost of the damage, which I may say we in the Works Department kept as low as possible. Can one imagine what it would be like for anyone who suffered with claustrophobia to be confined in solitary?

Castle Mill, Oldham, was a six-storey building of imposing red-brick construction, with a total floor space of some two hundred thousand square feet. Welton, F. W., was full-time at the mill premises during the complex building contract adaptations works, which included a goodly variety of items—propping of the floor beams to increase the carrying capacity of floors, which were now required to hold heavy stores, and work on the loading bay with platform, sprinkler system for fire fighting, hatchways inside the floor areas, offices, canteen, toilets, and other facilities common to many of the jobs for the Store Department. On the top floor, special dustproof, soundproof accommodation had to be devised for the Admiralty Signal Department.

The naval store officer in charge was F. G. V. Lott, a likeable character fifty years old. He was kept on his toes by the regular visits of the S.N.S.O., the superintending naval store officer, from Hazlemere, dispensing policy and ideas that Lott put into practice, keeping a bunk at the office, where he was often on call during the night.

One day while I was seeing Lott in his office on the second floor of the mill, my old black Ford 10HP Beetle was waiting patiently in the car park as I sipped tea made by Lott's ravishing young secretary. Outside the fog was descending, threatening to become a real Lancashire pea-souper. I was booked in at a commercial hotel at Rochdale for the night. Lott, from the kindness of his heart, insisted on accompanying me to the hotel, seven miles away from Oldham, just in case I should get into difficulties. At about five-thirty, as we tootled slowly along the road to Rochdale, the fog was thickening and Lott, fearing for our lives with the uncertain driving of his "chauffeur," suggested he get out of the car and lead the way on foot. The idea was a good one. The headlights now focused on the lone figure running like he was doing the marathon in the milky opaqueness of the fogbound night. He ran on steadily for one who had reached his half-century mark, his mackintosh and fair, thickly thatched bare head alternating in and out of the range of my masked headlights in the fog.

We continued along in this fashion for a quarter-hour; then, just as I was thinking about my high tea of steak and chips to come, I saw Lott sprawl out in front as my foot on the brake fair broke the cable. His heels had caught up in the bumper and upset his balance. Jumping up quickly, he called out, "I'm okay, old chap! Keep coming!" So, following my star, I arrived at the inn, where the pair of us, with the car, were more like a horse and trap as he led me through the archway and into the hotel yard, once used for stabling horses.

Memories of the Liverpool Works Office mean more to me than just work. They mean Lott and his like, Mather and Platt and Oldham and Rochdale, and steak and chips for high tea in commercial hotels whenever I had to stay the night somewhere in Lancashire, at a cost of around eight shillings and sixpence for bed and breakfast, with three and six for the high tea. They also mean frequent changes of steering gear as I rumbled over the cobblestones of Liverpool, St. Helens, or other towns in my very secondhand Beetle. They mean breaking down in the Mersey tunnel on two occasions.

On the first occasion the steering drag link fell off just after I had entered the tunnel at Birkenhead end, the car careering into the left side tunnel wall. The tunnel police came along and had me towed out for three times the tariff for the car. That amounted to four shillings and sixpence. The police at Liverpool were big and broad-minded, sympathetic people, and the tunnel police, too, I remember for their helpful attitude towards motorists.

On the second occasion I was stuck in the tunnel I was returning home from the Liverpool side after a long day's work in Lancashire when I ran out of juice in the middle, right under the Mersey, in the very structure Higginbotham had worked on before joining the Admiralty. This fact was of no help to me as I sat there in the slow lane at the bottom of the sag while the heavy mixed vehicular traffic scudded past me, waiting in my capsule for the rescue brigade to come along. The tunnel police saw me as they approached from the other direction. I saw them with a sense of relief; it seemed they were with me in a trice, and once again I was towed out, only this time they saw that I had half a gallon of petrol to send me on my way. All they suggested was that I, "Try to be more careful in future, sir."

I must put in a few flashes before we leave Liverpool area, which laid claim to two memorable periods in my early life, in the first of which at the university, I was conscious of not earning my own living, and this second episode, when those twinges of conscience were put to rights, as it was perhaps the busiest time of my life, in terms of personal effort.

There was a camaraderie in that part of the world in which travelling to work, in the hustle and the bustle, getting on with the war effort at work, being prepared to cope with emergencies in addition to working with no thought of extra payment, fire watching and fire fighting the results of the onslaught of incendiary bomb attacks on the large targets offered by such as the Liver and others on the dockline and in the city, with emergency water tanks dotted all around, supplied with flexibly jointed pipes laid up the streets for conveying water pumped from the Mersey as and when required, we on Merseyside knew we were all at it in a common cause.

In 1943, halfway through the four years I spent as an A.C.E., I applied for service abroad when vacancies arose for C.E.'s. Roger Higginbotham was still S.C.E. for the area. In fact, two of us volunteered, Harrison, older than I, being the other applicant for the same post involving promotion. Maybe the S.C.E. wanted to hang onto us, as we were both turning out a large volume of work, but a little bird informed us of the coldwater comments in the letter forwarding our submissions. Harrison was "Very experienced, but on the Architectural side," while Price was "capable of doing the work, but his extreme youth should be noted." Well, we each felt we were worth a move, but in my case, it wasn't to be for another two years. I seem to remember that Pitt the Younger was only twenty-four when he was made prime minister. I was twenty-six and had been in the game, young man and boy, since the age of nine.

Roger Higginbotham left the Liverpool office before I did, and he was replaced by none other than our old friend Joe Lumley, of Trecwn fame. Old Joe was a character whom I began to describe when telling the Trecwn tale, but in Liverpool he had to play a different role, as you can well imagine. It was a bit alien to his natural delight where work was concerned. Perhaps three years had elapsed since he was transferred from Trecwn to Admiralty Headquarters. When he came to join us as the new S.C.E., his style was completely different from that of his predecessor. I was just a young man in formative years with the Admiralty and had been able to adjust to the work, which was at a more junior level of responsibility than that of my old friend. My work required attention to innumerable details concerned with a miscellany of work scattered over a large area. Joe's job in charge of the whole works office responsible for the navy's needs demanded quickfire decisions all day long. Higginbotham had started the office and grown with it, running it in his own style. Joe was a heavy works man who was good at making solid decisions that would stand for all in an organisation to follow. In these terms he probably saw our area works as a hotch potch of all sorts. He delegated well, as always, letting people get on with their own work, but did not seem to have enough to occupy his mind once his out basket had been emptied, and spent his spare moments trying to catch sea gulls alighting on the balcony and then his windowsill. The gulls would often take the crumb from his hand but always evaded capture.

J. P. Lumley used to find time to tour the area by accompanying his roving personnel like myself, and I fell in line for most of these trips. Those occasions were valuable to me as experience, but the people we met as a rule were not what Joe could call his "opposite numbers," so his progresses made in this way were more like the visitations of the pope, learning at firsthand of the farflung people connected with his Admiralty service and duty.

"When I was at Trecwn, Price," Joe would say, "my stock was up. Since I have come to Liverpool, my stock has gone down."

It is interesting that Arthur Lang, my colleague of equal standing as an A.C.E., had the ear of old Joe quite a lot, both men being practical, down-to-earth human beings. When the Admiralty wanted three C.E.'s to fill the positions of divisional C.E.'s in the Far East, Lumley decided to put my name forward as one of the candidates. Lang reckoned he had tilted the decision in my favour, as he had previously said to Joe I should be a C.E. Be that as it may, my name was up for consideration in the latter half of 1944, and while we are awaiting the result let me tell you of things domestic.

During these weeks hoping for transfer and promotion abroad something traumatic had happened to our lives at home. Our young baby Linda (short for Lin's daughter), born on the third of July 1943, contracted whooping cough, which had been brought from school by our two boys, who were victims of an epidemic. Only three weeks previously I had been down to Pembrey for the week-end and the baby of fifteen months had been full of life in her pram, standing up and looking forward, supported on her hands and feet, bouncing up and down as I ran with her down the old dirt lane of the Furnace.

I learned the news when a telegram came to the office, "Linda very low," and my next flash was of my wife, mother-in-law, and baby in front of the fire in Cliff Cottage, the baby in a warm blanket, a bath of hot water alongside used to dip the child into when she turned blue with each convulsion. The performance of the sixth convulsion was demonstrated in front of me within minutes of my arrival. It seemed the doctor could only suggest this treatment, but it was more than I could bear. The doctor was along before another spasm occurred, and after I taxed him strongly on the need for the child to go to hospital he agreed. It was awful to remember our baby as she grasped my little finger in Llwyn-On hospital and to see her the next day on the mortuary slab in the same hospital. It seemed the specialist treating Linda had been administering antibiotics and explained that she had had too may convulsions for the treatment to save her.

Therefore, as I stood for a minute in the hospital mortuary on the twenty-sixth of October, 1944 it was the saddest moment of my life. In the lounge at Cliff Cottage on the day of the funeral, speaking privately to my father, I held up the tiny white coffin in my arms. My dad said I must get ready for the next disappointment in life.

Late 1944 saw me measured at Gieves, Liverpool, for my commander R.N.V.R. (Special Branch) uniform following my appointment to duty in the Far East. Uniforms were required to be worn out there, as personnel falling into the hands of the Japanese were liable to be shot as spies if they were in civvies. There followed a period of several weeks' anticipation of my sailing orders while I stayed at the Liverpool office, not doing much after I had handed over my own work to Halloway, the queries he had to raise reducing quickly as he took up the reins. C. P. Lowe had succeeded Lumley before my sailing orders came for the beginning of March 1945, and we did not see much of each other. "As one man falls, another takes his place," goes the old saying. Eventually, as I left the friends I had made in the Liver, Stanley Oxley said a few words and draped me in a dressing gown, brown with white polka dots, made of a silky fabric, a gift from everybody, those all around me.

Leaving Lime Street station for my few days' leave in Pembrey before sailing, I made what was to be a never-to-be-forgotten, never-to-be-fulfilled promise to meet up with my special friends after the war was over.

It was the end of another episode of my life. Once again my pen is inadequate to express the emptiness at my leaving Liverpool.

8. Upward and Away

Things were a bit strained emotionally when I returned to Pembrey from Liverpool in late February, as I had to sail for the Far East early in March 1945. It was still wartime, and going abroad to take on new responsibilities caused me to conjure up all sorts of possibilities in my mind. I had the notion it might be impossible to convey my exact whereabouts when writing home and so devised a code to give some idea of my location. The number of kisses at the end of a letter would represent numbers I had marked on a map to denote places like the Andamans, Nicobars, Ceylon, various parts of India, Burma, Malaya, Sumatra, the East Indies, and umpteen others. It was all quite unnecessary, and in any case, any screening of mail from P.O.W. camps would probably have completely obliterated such a naive device. There was something of a hiatus at the Cliff due to the fact that I had to leave the repair of the fireplace in the kitchen to Bebb, the plumber.

When the day came for me to sail, Lin accompanied me to Newport Mon. for me to join a ship sailing out east. Arrived at Newport railway station, we had lunch in an upstairs cafe overlooking the main street in the town centre and then took a taxi to the dock gates. Lin was not allowed through the gates and had to wait for the taxi to collect her on the way back. I can see her now as I looked through the back of the taxi, standing on the other side of the heavy iron gates, a slim figure in a Maid Marion hat with a feather, her gas mask slung by her side, feebly waving till the taxi twisted us out of each other's view.

The ship was the S.S. *Strategist,* a six-thousand-ton cargo boat with a dozen cabins, according to the stevedore, with his hand on my black officer's trunk before it was yanked aboard by the ship's crane and dropped into the hold. The ship was owned by T. and J. Harrison and had been registered at Liverpool, my city of destiny. On board, I discovered that sixty passengers for the Far East were being accommodated in what were twelve single cabins in peacetime. We were now in the care of the master, Alfred Gray Peterkin, and his 141 members of the crew.

There were six bunks in my cabin, and I occupied the top bunk by the door as I went in, two of my cabinmates being also of the Civil Engineers' Department and proceeding under orders on the same basis as myself, to serve as divisional civil engineers, J. E. Newton and N. D. Carter, the latter having attained his degree at Liverpool in the year 1937.

That night I felt no romance at the prospect of the service abroad ahead of me, apart from which I had a stinking cold. On the ides of March the ship pulled out of Newport Docks and sailed down the Bristol Channel and along the South Wales coast to anchor in Milford Haven harbour, where a convoy of ships was being assembled prior to sailing under escort of the navy.

Three days later we were corkscrewing along in the Bay of Biscay, which caused my insides to succumb, and I deposited a couple of partially digested meals over the side. It seems to be my nature to fight against seasickness for about three days, then let it all hang out before settling down to enjoy a voyage. One benefit I enjoyed was the disappearance of my cold, and with that came a raising of spirits. Occasionally, lying in our bunks, we would hear bumps in the night as depth charges were dropped by our escorts to flush out or deter any U-boat with designs on our fifty-four strong merchant fleet of vessels.

At Gibraltar, the convoy broke up and a few of the number turned left, as it were, to enter the Mediterranean and head east, the others proceeding to their own destinations southward or westward. The blue of the Mediterranean is something to be experienced at sea, and this is something war cannot alter, also the sight of a school of dolphins at play round the ship. Our vessel was probably doing in excess of fifteen knots, and yet these beautiful creatures would swim round the ship with the greatest of ease, their bodies glistening and, at night, making wonderful shimmering shapes in the luminescent surface of the water.

Off Morocco, one of the passengers from our cabin, going out to a tea plantation in Ceylon, was given to singing a popular tune, "Oh, I'm Off on the road to Morocco," the lighthearted ballad wafting over the deck putting many of us into good spirits. We somehow felt a bit safer now that we had the Atlantic behind us.

Through the Suez, with banks both sides like a big version of the Manchester ship canal, we wondered how the chief cook was getting along in Port Said after being put ashore with bladder trouble. It was April 2, the heat mounting as we headed south in the Red Sea with Aden at the southern end of the Gulf of Aden, our next port of call. The ship was taking fuel aboard while we three fledgling promotees from the admiralty were allowed ashore to see the town, hot, foreign, and foreboding, with its steep, hilly background behind the main street parallel to the shoreline of the harbour. We were promenading along this street with nothing to do, trying to decide which of the eating houses looked least undesirable, having been spoiled quite a bit on the ship.

We were silent for a while as we walked along the street, but my mind was active thinking over the incentives required in the psychology of winning a war. The old adage "An army marches on its stomach" also applies to the navy, I mused. Good food and a tot of rum are essential for morale, or there'd be no ruddy navy. Patriotism is about fighting for one's country and at various times emerges in high relief. It is about sacrifice and about camaraderie in sharing the common cause, but it goes better if the body is kept on a high with the brain topped up with oxygen from rich, red blood, sustaining the will to win the cause. The need is heightened on board ship, where one is bound for unknown destinations on a floating island with no enemy in one's own personal gunsights.

Civilians in the U.K. had to survive with perhaps a quarter of the rations of service personnel, in terms of butter, cheese, and meat, but there was no complaint of this because of the regard people had for the lads in the front lines, though God knows everyone was in the front lines. After a blitz, looking around, one's thankfulness to be spared preceded the revulsion at seeing the dead. Folks were kept going by the will to win inculcated by leaders whose job it was to whip up the spirit of patriotism and keep it burning strong and sure in the knowledge that Britain would emerge triumphant. We were at one with the knowledge that Germany had been arming to the teeth, "guns, not butter," and the enemy would therefore have even less to eat than ourselves which for those not involved in the black market rackets was damned little. So back home people were kept going with incentives, part of the £13 million a day we were spending on waging the war, letting off steam when the nation at last blew its top with the dastardly deeds of the dreaded dictators on our doorstep.

When mills in Lancashire were replaced by naval establishments, there were built-in comforts, like modern waterborne toilets and canteens, to a standard never available before the war, and the novelty of "music while you work" coming over the tannoy systems, punctuated by the laughable, if funereal, voice of William Joyce, alias Lord Haw Haw, "Jaremany calling, Jaremany calling," brought fun into the world drenched in the carnage of bombs, high explosives, incendiary bombs, hard hats, emergency water tanks, and bloody chaos after each admonition by the enemy as wave after wave of enemy assailants of the air sought to destroy our patriotic will. In the holocaust the spirit of Christ was interpreted by each side to suit themselves, "turning the other cheek" being eclipsed by "fighting the good fight," as Neville Chamberlain's paternally patient view of the foe had suffered a shock and the nation's earthly saviour, Winston Churchill, sounded the clarion call in language all could understand, with his Anglo-Saxon talk, his V sign, and the Morse beat transmitted over the wireless at news and other intervals of time.

"What are we going to do?" Newton broke our reverie of silence.

"We had better try somewhere," we answered in unison.

Nick Carter was a tall man with straight stature, angular jaw, and slightly auburn hair cropped tidily back and sides, with just the suspicion of freckles on his bespectacled physiognomy. He was twenty-nine. Newton, five-foot-six, a lightweight, fair-complexioned, thirty-one-year-old, already a bit thin on top, mounted the steps of an eating establishment that crowded onto the main road as though being pushed from behind by the high hillside terrain. Nick and I followed.

Once inside the bare room with spartan tables, we were not titillated by the sight of the platters on tables round about us. There was too much oil associated with the dishes, both the salads and the fried fare, and Newton came out with, "We'll just have eggs. If we have them boiled there won't be anything wrong with them."

I was never one for eggs. "I'll just have one," I declared. If I had more than one at home, the albumin content always caused me to overheat and break out in spots, but I decided to go along with the others as we sipped our beers poured into glasses from cans with screw tops. When the eggs came they were bantam size and each one barely a mouthful. We were glad to return to the *Strategist* before it weighed anchor to sail off into the red heat of the Gulf of Aden to make the last leg of the voyage with the long trip across the Indian Ocean to Colombo. There were no particular incidents as we sailed steadily over that member of the seven seas, just an occasional dhow heading for the islands and one liner on the horizon making its way in the opposite direction to ourselves.

On Wednesday, the eighteenth of April, we were tied up in Colombo harbour. It is not the place to hang about, not if you have just come in from the U.K. Your thick blood is trying its best to thin down, and you are in the high nineties, sweating like hades, with humidity oodles percent. For many fair-skinned types, prickly heat is inevitable. We hung around for a day before being released ashore. Our trip had taken nearly five weeks, and it was nice to step on land, like drunken men until our sea legs grew accustomed to being on terra firma.

Up in the rest house at Mount Lavinia, the exotic nature of the country struck me forcibly. This was some sort of paradise island. What would those two young girls who had come with the ship think of their new home? They were from Swansea and had married two Ceylonese students who had qualified at the University College of Wales, Swansea, with its lovely setting in Singleton Park. On the way over, both these young men had spoken of Wales being very similar to their own country, with a similar lilting tongue of the people, but the similarity they saw was possibly because they had just spent a solid three years in an active university life, developing mentally and physically, the sap rising in their young bodies, so that really, they had identified their host country with the girls they had taken to their beds in marriage.

In the early hours of the morning, as day began to break, I lay back on a hard, comfortable mattress, looking through a mosquito net with the large-bladed, overhead fan whirring smoothly, stirring the air in the room, and listened to the melodic tuneful call of a songbird waking a sleepy world with notes to grace this entrancing tropical island. Repetitive strains of eight notes to the bar, enough to lift a young Mozart, Schubert, Liszt, or McCartney to the ecstasy of compulsive composition with sound to match the exotic flowers and verdures of the town parks seen yesterday.

Then, after breakfast, with pawpaw on the menu, my first taste of that tropical fruit, I went over to the offices of the Assistant Civil Engineer in Chief, South East Asia, R. Morton, for information as to my posting to the Indian subcontinent. Carter, Newton, and I were interviewed for suitably allocating the three posts, one at Administrative Headquarters of S.C.E. India at New Delhi, one at Bombay in the office of the Superintending Civil India North, and the other one at Madras for S.C.E. India South.

It was decided that Newton would go to Delhi, Carter to Madras, and I would be posted to Bombay. Newton would fly and we other two would travel by rail.

As I was swimming in the sea near Colombo my left ear suddenly blocked solid. The discomfort was acute as I tried to shake my head free of the blockage. The young service doctor, a lieutenant R.N.V.R., did not need a special optical aid to see the sand blocking my appendage. He got to work with a syringe, and out in the kidney-shaped bowl came a dollop of yellow sand. It was some of the stuff held in suspension in the sea as the waves splashed on the golden sandy beach and returned with strong undertow.

It was time to say good-bye to folk at the rest house, to S.C.E. Colombo, T.A (Tammy) Burnside, known previously to me as C.E. 1 at the royal naval armament depot, Trecwn, as Nick Carter, and I boarded the afternoon train from Colombo to the north. There was sufficient daylight remaining for us to take in with appreciation a cross section of the country either side of the line—the lush greenery, the palms, the village houses, some with concrete block walls and red roofs, others of palm frond cladding. Local trains coming into Colombo had bodies clinging to the outsides like limpets.

I was to spend the next five nights and four days travelling overland by train. The route lay via Talaimannar, at the northern tip of Ceylon, across to Danush Khodi on the southern coast of India, then on to

Trichinopoly, Pondicherry, and Madras and across the subcontinent to Bombay. The wonder of that journey is worthy of a separate tale, but I am anxious to get us to the other end of the journey and so will confine my description to a few instantaneous flashes. The Germans have a word for them—*Augenblicken,* and we'll have ten of them to meet our purpose.

Augenblick no. 1: A mêlée of people during the transfer across the straits between Ceylon and India.

Augenblick no. 2: Nick Carter and I have tea in the green-walled, high-ceilinged buffet restaurant of the station at Trichinopoly.

Augenblick no. 3: The catering of Spencers and another firm still on the tip of my tongue after nearly four decades.

Augenblick no. 4: Nick Carter leaving at Madras station, to take up his post as C.E. in charge of the vast naval plant depot at Avadi, near Madras.

Augenblick no. 5: The platforms snying with all forms of Indian travellers with aluminum food containers in many tiers and baskets of fruit. Extremes of society from Brahmins, high caste, upright, superior, clean white apparel, carmine-coloured spot on the forehead, and, by contrast, the most noble of all God's people in the land, the lowly beggars chanting incessantly, "Baksheesh, sahib."

Augenblick no. 6: The train compartment with its four sets of window covers, glass, louvres, mesh, and shutters to suit the prevailing circumstances.

Augenblick no. 7: My companion after Madras, a captain in the Indian army, at home in the leather style appointment of the compartment, yawning while I remained agog.

Augenblick no. 8: The vast terrain of brown, parched, earth, of dry nullahs, of deep, chasmic valleys spanned by high viaducts but devoid of water.

Augenblick no. 9: Rolling into the station at Bombay and seeing a tall dark-haired, white-faced young Englishman waiting in the throng to meet a compatriot due on the train from Madras.

Augenblick no. 10: As the train creaked to a halt, large canvas hold-all bag in hand, I stepped out as he greeted me with 99 percent certainty: "Mr. Price?"

"That's me," I confirmed.

"I'm Wilcox. The S.C.E. asked me to meet you and take you for a shower and a bite before coming to the office."

Then we were esconced in a four-by-four station wagon, a boxlike transport with a swarthy driver at the wheel.

"Back to my place, Rodriguez," came the word of command. "Did you have a good trip?" was a question to me.

By now, eighty-thirty in the morning, I had spent five nights and four days rumbling all the way up from Colombo and was grateful for the shower and change of clothing into my shorts and shirt of khaki drill. The kit I had bought in Liverpool had included both white and khaki shorts, shirts, and stockings, also three pairs of stout brown shoes of exactly the same last. However, I wore no uniform epaulettes, the need for uniform having disappeared in Bombay. Some of our department still besported their ranks, but I was to find the freedom of my civvie status more useful. Showered and breakfasted, I heard Wilcox, a clerical officer, declare, "We'll have to report to the office now, Mr. Price."

Ewart House, Bombay, stood on the corner of an intersection of streets. On the first floor Wilcox guided me through main doors, through inside swing doors, then into a large open room full of drawing desks and people, mostly heads down to their jobs but some looking up as we moved quickly down the centre of the room to the far end. On the right was a corner partitioned off with a light-framed access door, which we ignored, but straight on was a solid door set in a solid wall, and in we went.

Seated with back to the large open window behind him was a fresh-faced Englishman in his thirties with dark hair parted and brushed down flat, whimsical smile atop the three wavy navy stripes on the shoulders of his snow-white shirt. He was A. G. Allnutt, the S.C.E. India North. Wilcox withdrew.

"Ah, my deputy," said the energetic young man in charge of the navy Works Office. The word *deputy* had the freshness of the business world, maybe demonstrating a desire by Allnutt to withstand the pressures of the dyed-in-the-wool attitudes of permanent service personnel. At any rate, I had never heard the word used in the nearly six years of my temporary service with the Admiralty. "My deputy, you've taken your time coming. I've been waiting for someone else for ages."

To be fair, the job was a big one for a youngish man, who had had to put together an organisation for the department, from scratch, to cater to the needs of the navy so far as works were required in the North India part of the eastern war theatre, thousands of miles from the U.K. and based on the assistant civil engineer in chief in Colombo, who was responsible for overall works approval, though administrative approvals had to be sought from Delhi, where S.C.E. India liaised with government. The existing "deputy" to S.C.E. India North was in hospital under the strain of the task in the heat of the Indian sun, which had resulted in a nervous breakdown.

"I have to go off to Calcutta to see Broadway, my A.C.E. over there," declared Allnutt. "You'll have to take over for a week."

Naval Works staff picture inside Ewart House, Bombay, 1945. Author is fifth from right, seated in the front row.

"When are you going?" I queried.

"Midday," said Allnutt, which conjured up in my mind shades of my Liverpool "Campbell" experience. Imagine a temporary civil engineer, just promoted, officially with effect from the minute I reported at the Bombay office, after six weeks in the doldrums, making my journey out like some nomadic tribesman in the desert, arriving at the Bombay office and having to take it on with some two and a half hours' notice. Well, I had always been one for a challenge and here was something to get my teeth into.

I sat by the S.C.E. as he scribbled some names onto bits of paper. Commander Wailes was a liaison man with the Admiral's office. Major Young liaised for the army, with their headquarters at Colaba. A few other names were dropped during the course of my brief briefing, including an indication of the split up of the territory. A few short epithets of advice poured out in between pieces of paper and files being signed or initialed by the busy executive.

"We'll get over to Firdaus for a quick lunch," he said. "I'm flying off straightway."

Well, Firdaus was his flat accommodation, and to get there we were driven in the office car, a white Ford Pilot, thirty-horsepower and of American manufacture.

The meal was a rush, but this suited the S.C.E., who was keen to get away before something prevented his apparently vital trip. The rush also suited me, as I had visions of a week of sheer hell, trying to make a go of a job that, for me, had as much difficulty factor as Churchill's had in number 10, probably more, W.S.C. having greater confidence than J. P. I was soon returning to the office, through the drawing office with its centre aisle like the hallowed walk of a church, up to the solid door of the S.C.E. office, closing it behind me and sitting on the swivel chair with, to put it mildly, mixed feelings concerning how to succeed as proxy S.C.E. India North by really trying.

From the left-hand flapped breast pocket of my khaki bush shirt I pulled out a packet of cigs and put one in my mouth. From the right hand flap my finger and thumb extracted a box of matches. I lit up and puffed gently as I looked at the array of baskets at the far side of the spacious desk, the desk where the buck stopped.

In came a young boy wearing one of those funny little forage cap–type of hats worn by Hindus and attired in Gandhi-like clothes and walking on cheap leather sandal-type shoes at the bottom of his spindly legs. He carried a couple of brown paper-backed files, each one inch thick, doubtless containing the results of months of toil and sweat, put them into the in tray, and then went out again.

A hefty man of fifty barged into the office, wanting to see Allnutt. "I'm Munro, C.E. Dredging," he announced.

Well, here's a turn up for the books, I thought. *At least there's a fellow engineer around to talk over matters.*

My hopes were quickly dashed as the Englishman explained, "I'm working on the dredging, which is being carried out by the Tilbury Dredging Company." It seemed he reported direct to the S.C.E. and had no connection with any other business of the office. He went out.

In came the young Indian boy again . . . and again . . . and again, and the files began to pile up in the in tray until they overflowed onto the desk space alongside. I stubbed my cigarette in the ashtray. "Taid, give me strength," I uttered. Then as the boy came in again, I came to life with an inspired determination.

"Boy, who are you?" I asked.

"I am Peon, sah."

"Peon who?"

"No, sah, I am peon who bring the papers, sah."

"Well, peon, who is your boss?"

"Mr. Apathuri, sah."

"Ask Mr. Apathuri to come in to see me please."

He went out. The door opened as I bid the person tapping on it to come in, a poor man's version of the old Aga Khan, the heavy one.

"You are Mr. Apathuri?"

"Yes, 'essir."

"And what do you do?"

"I am chief clerk, 'essir, chief Indian clerk, 'essir."

"Do you know about these files?"

"No, 'essir."

"Who knows about them?"

"The chief clerk, 'essir, chief European clerk."

"And what is his name?"

"Mr. Ballard, 'essir. He is D.H.C.O., 'essir, the departmental higher clerical officer, 'essir."

"Would you please ask the D.H.C.O. to come in to see me?"

"Won't you wish to get on with your work, 'essir? There is much to do." He was looking with disdain as the peon came in with still more files, now being stacked up in piles on the desk where space could be found.

"Right now, please," I insisted.

He went out.

There was a tap on the door, and in came an iron-grey-haired man in his forties, moving a bit like Groucho Marx, with the determination of a long distance runner, his heavy horn-rimmed spectacles held in his right hand, his left hand rubbing his conscientious eyes.

"Am I glad to see you," he opened. "We've been getting a bit snowed under with no C.E. Bombay."

"Well now, Mr. Ballard, I wonder if you would help me, I wish to attend to these few papers." The word *few* was not emphasised but used normally as I introduced my objective of surfacing from under the sea of bumff.

He took a seat by my side and gave me a wry look.

I picked up a file from the desk. "Which A.C.E.'s area is this?"

"Let me see. That's Chembur Camp. That will be Mr. Abell," he told me.

I put the file at the right hand near corner of the desk.

Ballard sat like a cat on hot bricks as I deliberately went through each file and allocated the whole lot into four separate piles to represent the four divisions for the navy works in the Bombay area. Then I thanked him for his kindness and left him to get back to his backlog of work he kept talking about. As he went out of the door I said, "Oh, and by the way, send in Mr. Abell."

Young Abell came in, fresh-faced, *Twenty-four at most,* I thought. He exuded good humour, the bubbly enthusiasm of youth. I felt better as he walked into the room and thought of Robert Young!

"Take a seat," I said.

He sat at the other side of the desk.

"No. Bring your seat and sit here."

He sat at my left while I looked at the first file. He filled me in with anecdotes of Chembur Camp. The other files made equally fascinating short stories of the navy at work in the mystical East. As each file went over to Abell, I made a one-line jotting on a foolscap sheet of paper.

Then I asked him to take away the files and would he please send in Bailey or Blount or Lovelock, whoever was available. By about three o'clock there was a tap on my door. Apathuri came in with some papers. He looked at the large empty desk as I stood smoking and looking out the window, Joe Lumley style.

"'Ee's magic," he said. I was lifted by the compliment, thankful for the richness of experience in my early life, thankful for Apathuri, even more so as I coaxed out of him that he was a father with many, many children being raised on the meagre wage of an Indian clerk, his fat form moving round with the shape and authority of the Aga Khan but his reward on earth being many, many children and his wealth not measured in gold or diamonds.

That first night Wilcox dropped me off at the Soona Mahal—a

multistorey block used by the British navy, one of a line of similar blocks along the Crescent Road fronting the Indian Ocean and stretching from Colaba at one end to Malabar Hill, the sacred garden of rest of the Parsis, where vultures eat the bodies of the newly departed souls.

The Soona Mahal cost me eight chips a day, which, being interpreted, meant eight rupees or twelve shillings of English money. This had to be paid out of my expatriation allowance on which I aimed to live while away from home, three-quarters of my basic pay being left in the U.K. for my family and future needs. In my small room with bathroom, I had the dubious pleasure of the company of a naval lieutenant who put me in mind of my boyhood picture of the Ancient Mariner. He chuntered away lying on his charpoy as I lay on mine, drinking tea brought in by a young Hindu domestic by the name of Ranji.

These modern multistorey buildings were probably the forerunners of the tower blocks and were already beginning to reveal their adverse qualities. Outside our room, across the corridor, was a small cell-like room where Ranji and his ilk sat, messenger fashion. Just at the side of this den was a window overlooking the light well in the centre of the whole building, and at the bottom, several floors below my level, the refuse of days had piled up feet thick with a pinkish pawpaw shade, colourful from up on high but sadly in need of collection.

"When the war's over, there's going to be a slump, and I shall be well off," confided my fellow bore, the lieutenant. "My pension will be adequate to see me all right for the rest of my bachelor life."

It was an environment designed to make me look forward to my return to the office, and the next day I was down to the executive desk with a will. My mind was uncluttered by too many prejudices or facts because I had been plummeted into the situation. Peter must have felt the same way when he was sanctified and given the job of custodian at the Golden Gates. So on Tuesday, May 1, I was able to look more carefully at the papers as they came in. It was immediately obvious to me that it was normally taking three days from the receipt of mail as date stamped in the office until the communications arrived on the oracle's desk for attention. Churchill had pointed the way—"Action this day"—but the Indian day must have been a long one. I pushed the button in the desk and in came the peon, who fetched Apathuri, who fetched Ballard.

"Why are these letters taking so long to get to me?" I asked.

Ballard looked at me like I had just exposed myself to the king. "Well, there's a lot of research goes into the preparation of my minutes," he said. He was referring to the small billet-doux, some five inches by seven, that were prepared and attached to all official communications.

I looked at the typical specimen. Clearly his department and he had gone to a good deal of trouble to spell out the administrative problems involved in dealing with each item. In particular, fleet orders were always being issued by the Admiralty, rather like the number of minibudgets introduced by a certain postwar chancellor, causing dire dilemma to Inland Revenue personnel, so the A.F.O.'s threatened all serious-minded departmental higher clerical officers in the Admiralty service, the "permanent bastards," as they were called, the term including all permanent service personnel. Dire consequences if orders were not complied with.

Well, I for one took a different line. In the short time I had spoken to Allnutt the previous day, it was clear he had been trying his best to breathe fresh air into the tired lungs of Admiralty life, and I decided to go one step further, not even bothering to give a second thought to what the S.C.E. in absentia would say, as I rapped out, in as kindly a fashion as possible, an instruction to the D.H.C.O.: "We will take these off," and from each file I ripped off the minutes, while Ballard's eyes popped out of his head.

"You can't do that," he said.

But as I already had, he was at a loss how to deal with such unforgivable behaviour. However, being in reality of very sound Cockney common sense, he soon fell into the idea of cutting his work by half at a stroke. "I'll have the mail on your desk early after arrival." He fell in with my request 100 percent, probably fed to the eyeteeth with the bloody A.F.O.'s. Besides, the man in charge always had to be aware of standing instructions for himself and be on the ball when coming to a decision.

The first week went quickly after that. Quick trips around the area, Bombay Island, with its manifold works executed by contractors, by direct Indian labour, 80 percent by women with head pans, by the PWD agency or by the Bombay Port Trust. The drill was, in fact, much the same as I had known in my four years at Liverpool, but now I had more clout, and by the end of the first week in Bombay, I was thoroughly enjoying my new life.

On the Monday in came the S.C.E. and I moved away from his empty desk to let him sit down. My list of occurrences on foolscap was conveyed to him, and he gave me a friendly indication of what was going on in Calcutta, not that it was of my concern officially. I mentioned that Wailes had been helpful to me when he came in to visit the office during the past week. Allnutt and he had a very good working relationship, possibly because of the extent to which they had a mutual dependency. There was one matter he kept back to be discussed with the S.C.E. himself, as the man "on seat" at the Works Office.

Now, I thought to myself, life should be relatively straightforward, looking after the day-to-day works matters in the Bombay works area, my own nomination being Divisional Civil Engineer, Bombay.

"You'll find your office first on the left," said the S.C.E., emptying his briefcase of Calcutta papers.

I went out, in through the lightweight door in the lightweight partitioned off corner of the big open room. Immediately inside the door of the seven-foot-high partitioning that passed for an office wall was a wooden desk, large, with an assortment of files and papers pending, including quite a few sheets of old calculations strewn willy nilly.

I was just going through them, the tortured work of my predecessor's mind in the period immediately preceding his hospitalisation, when in walked a man and pointed at me. In a voice like an actor making a stage entrance, he said, "You are sitting in my chair." I felt like he was father bear and I was eating his porridge.

Yes, it was obviously the good man I was relieving. The psychological solution of the situation was instinctive. I rose from the chair. "Thank God you are here," I declared. "I am just about up to here." And with a mock gesture I put my stiff hand at the level of the top of my head.

"Oh, no, no, no," he said. "I was only pretending. I am being repatriated."

There were several chaps I knew in the eight years I served in the Admiralty who succumbed to overstress. Maybe these men were the real heroes, men who lost themselves in the unknown darkness beyond the limit of their own endurance, men who worried themselves sick, men who sometimes worked on and on, though in a fever in a foreign land. What unknown darkness of the mind does a man walk through when he sees Nissen huts in the place of motorcars in the streets of Bombay. I wondered if I would be taxed to that extent.

We shook hands warmly. He went out of my life as he came in—suddenly, all in the space of minutes. His legacy of struggle at his desk lay behind him, taking on a new meaning for me, his thoughts coming quickly to life before my very eyes, and by the afternoon I was once more up to date in the office. There were now no skeletons left in the cupboard by some impersonal predecessor.

Now I really must press on, because we are only eight days in Bombay. It is the seventh of May, and I have to tell you of about a year in that internationally famous city.

There were outside Bombay Island places like Thana, where the Admiralty had a big store depot comprising Bellman hangars and also camps and hutments all monsoon-drained and mosquito-proofed. To these more remote establishments A.C.E. Bailey travelled relentlessly in his station wagon, taking me with him from time to time, plans rolled up in the back, bowling along on the highway without ever applying the brakes, dodging the mêlée of pedestrians, bullock carts, garries, and all sorts parting in front of us or being bypassed according to the needs of the most skilful driver I have ever known and the second most frightening. (The accolade for fear creation in his passengers in my lifetime goes to Walsh, site clearance contractor, doing ninety miles an hour in the country lanes while putting his 4.2 litre Jag to the test.)

Abell was stationed for a time at Chembur Camp, where he was overseeing the modernisation thereof, the monsoon drains, the kitchens, the domestic sewerage, and sundry building modifications. On a visit to the camp one day I was sickened by the sight of the preparation of food in the kitchen. Huge fish lay out under the verandah outside the gauze-covered door propped open for air to get in to the hot kitchen with furnace heat from the solid fuel stoves. Sailors were chopping the flies swarming on the dead marine creatures into the flesh, making it impossible to know whether it was fish or fly blood reddening the steaks. My sense of hygiene was greatly offended as I relayed my observations to the camp commander. It was an impossible situation for the camp commander or young Abell to rectify. I had merely drawn attention to the need to keep the mosquito doors closed. No one could enforce that one, given the camp facilities. Boiling of the fish would hopefully kill the flyblow, but who could be certain of that? Often in life it is not a case of this and that; it is a case of this or that, and this was such a case.

The Bombay Anglican cathedral, a Methodist chapel, and the military hospital at Colaba all had the doubtful privilege of visits from your author.

The cathedral was a splendid edifice, but for friendship and song I was moved to join a little Methodist chapel community, and that helped to speed along the time of my stay in Bombay.

It was strange how I found myself in hospital. Never in all my life had vaccination, the shield against smallpox, taken. It was the same when the medical officer tried to do it at Liverpool, twice without success. It was the same again at Bombay, when they had another go at me and with no result; another scratch of the lymph was tried a few days later. Then the first scratch took and after it the second, each spot on my arm developing the normal type of inflammation and sore, but after the second positive result, my temperature began to rise with vaccine fever. The naval doctor gave me a note to go into the military hospital at Colaba, the tip of the Bombay peninsula.

A transport dropped me off at the gate, and I reported at the office carrying my leather week-end case, the one with J.P. on it, my twenty-first birthday present from my parents. There I had to fill in a form, though it was difficult to make out the questions in my state of health. Then I had to make my way over through a yard to a ward on the ground floor, taking up a pew in the porch affair outside the main door. A nurse took my form, stuck a thermometer in my mouth, and went off. When she returned to look at the thermometer, I could just make her out as she jumped about a foot in the air and ran into the ward. I was flat on my back as men in white coats took me inside the ward. Big white pills about the size of halfpennies were suddenly being forced down my throat, and soon I was in dreamland. Then I have memories of waking, swallowing pills, sleeping, and the whole process going on repetitively for some days before my temperature subsided enough to take notice of the pleasant English environment of the hospital: the cleanliness, the femininity, the lovely food, the Cadbury's chocolate and other brands known to me. I was sorry to leave the hospital, but upon my word, my weight was down at least ten pounds than previously noted, and even before going into hospital I had been dehydrated considerably.

There was another association I had with a hospital, but not as a patient. Elphinstone School in Bombay was being used as a naval hospital, with patients packed out on to the outside verandahs. *Chieks* of bamboo cane and matting were required to shield off the imminent monsoon, which was due in mid-June. Contractor wallah, "Mr. Contractor," had the contract to fabricate and erect these contrivances, and he was not conspicuous by his presence or, rather, by his *chieks*.

Whenever I looked into the building with its patients full to over-flowing, my eye looked at the open verandahs on each floor, and as soon as Mr. Contractor came over to the office I would have a go at him about the matter. He would nod in the way they do in that part of the world, assuring you with a shake of the head from side to side, which seems more like a "no" than "yes."

"They are being made," said this man, with irritating phlegm. "They will be up in time for the monsoon season."

When in Lancashire, I had seen steeplejacks perform miracles with economic scaffolding on chimneys and high buildings, but these merchants in Bombay seemed to be carrying brinkmanship a bit too far. With only two days to go to the monsoon date, the fifteenth of June, the contractor wallah's crew began. Up went the spidery network of bamboo-poled, jute-tied scaffolding, and up went the *chieks*, the barefoot men climbing up nimbly to secure them firmly in place.

The first day of the monsoon 9.7 inches of rain fell on Bombay, the outermost patients safely protected in the relative cool day comfort of the balconies outside the main wards.

A basement of the Bank of India Building, under construction, was in the possession of the navy, having been requisitioned for their purposes, but as it filled up overnight at the outbreak of the monsoon, it was just as quickly derequisitioned.

Two feet of rain fell in the first week of the monsoon. It made up for the nine months' dry season in those parts, when the sun becomes unbearably monotonous.

Some of the most pleasant times I remember in India were the hours spent sailing in Bombay harbour with the S.C.E. in his dinghy. He manoeuvred the small craft with great skill and taught me a thing or two about the art of sailing.

Captain Farquhar, the commanding officer of HMS *Braganza,* is worth a mention. What a man he was. He was as lovable as Long John Silver of Treasure Island, but instead of a parrot he had a terrier and instead of a wooden leg he had a walking stick, which he used almost as often as Jomo Kenyatta did his flyswatter, laying about any matelots who crossed his path.

Braganza, a famous Portuguese/Indian name if ever there was, was established in the Wodehouse Barracks when the Indian army moved out. It lay alongside the main road and at the one end was some eighteen feet below the road. One day at the office in Ewart House, I was tackled by Lovelock, A.C.E., whose area included the shore establishment in question.

"Price sahib," he began. (He affectionately called me sahib.) "Price sahib, we have a problem at Wodehouse Barracks. Let me show you what I mean." He spread out the plan on my desk. "It is a question of the drainage on the site," he began. It always is.

The difficulty was that the site was planned to drain its domestic sewage into a corner where the road and main sewer were well above the barracks, a bloomer of army planning. The new drainage was being laid to replace the old thunder boxes of the barracks, and it only remained for the outlet pipe to be linked up with the public sewer. Now we had hit a snag. The sewage could hardly be expected to flow uphill.

"What about the road at the other side of the camp?" I asked my A.C.E.

"I'll go and see the municipal people and have a think about it," said my worthy assistant.

The next day he came back with the glimmer of an idea. "It seems that the road sewer of the other corner," he stuck his finger on the plan at the adjacent corner of the barracks site, "the sewer there is just low enough, but it is surcharged all day so we can only have access to it during the small hours of the morning, between midnight and 6:00 A.M. We would need a storage tank for the daytime effluent accumulation. I've sketched out a layout, but it's just a thought. We haven't a hope in hell of getting a couple of valves to control the flow at each end of the chamber."

"It so happens that we *do* have two valves, two eight-inch-diameter ones surplus from the new oil jetty," I was able to assure Lovelock. (My Taid had been looking after my welfare once again.) That day saw the two of us going to sell the idea to the old captain. We would need his help. Someone would have to be operate the valves manually.

Lovelock was keen that I should be with him. "Better if you come along, Price sahib. We cannot afford to be obstructed by old Farquhar's piebald liver." We asked the naval health officer to come along to give more credence to the proposition afoot.

Captain Farquhar was a veritable character, an exhibitionist, extroverted and ebullient, the type who enjoys an audience. When he understood that the army had made a bloomer and that the navy now had the insuperable task of manning the necessary mechanisms from then on, the captain was delighted. The health officer's insistence that a high vent sticking thirty feet up in the air like a ship's funnel raised no objection from the old sea dog.

"You mean I'll need to have personnel on watch from eight bells, midnight, to four bells." It was obvious the Captain was revelling in the prospect of adding to the nautical flavour of HMS *Braganza,* which already had a quarter deck and now would have the semblance of an engine room, given enough imagination. We came away from the old salt infected by his exuberant enthusiasm.

It was not all work at Bombay.

At the Brabourne Stadium, I sat in the stands watching the 1945 Australian Services cricket touring team playing an Indian eleven in early November. Among the players were India's Western Zone's openers Merchant and Mankad, high-scoring Modi, and all-rounder Hazare and Australia's stalwarts, like Hassett, Carmody, and Pettifer, with Keith Miller, the famous all-rounder, then in fine form, as exciting as Ian Botham today.

During the game there came a young Englishman with friends, and he sat in the stands in the next seat to me. It turned out he was a son of the Pilkington family of St. Helens Glassworks fame.

Our Mr. Old, a storehouseman currently deputed to sort out prefabricated parts, and nuts and bolts for storage shed erections, had been singing the praises of the Bombay races and was an avid follower. Many of the locals were inveterate if not professional racegoers and would plunge heavily as fortunes rose and fell out of all proportion to weekly income. I could count on one hand the number of times I had been to the races in my lifetime, but on this occasion I was looking at the paper and saw that Tommy Burns, an English jockey, was riding in seven of the nine races. I decided to go along and try my luck by following Burns through the card. On the Saturday afternoon I banked on arriving in time for the third race, the three o'clock, which was Tommy's first race out of the nine that afternoon.

Frank Webber, our quantity surveyor, and I arrived just too late for this, and I was mortified to see that Tommy had just come up at 195 chips to 10. Well, I followed him belatedly, but for the sake of not being a few minutes earlier I was destined to have my usual lesson not to bet on horses, though I seem to remember that Burns did have moderate success in the later races, by which time I had switched my allegiance elsewhere.

Frank Webber and I used to spend some evenings playing billiards at the Majestic Hotel, not far from the Gateway of India, a magnificent pair of properties by which to remember Bombay. The Gateway offered a fitting stonelike portal to enter the great city from the harbour, with its solid squat architecture, nearly a hundred feet high and with width just right to put one in mind of the Arc de Triomphe in Paris or the Marble Arch in London, each in its way of similar value to its citizens. The Gothic arches, the double line of cornices high up, and the four round towers with cupolas topping off the corners of an imposing high centrepiece, distinctive from the distance, made for a never-to-be-forgotten site.

The Majestic, standing off the bifurcation of the Colaba Causeway and Wodehouse Road, boasted of architecture in brick and stone, soft, rounded pilaster-type lines, turrets, arched windows, and a large entrance below a Gothic arch, a fitting building to stand in the world-famous city.

Inside the billiard room, Frank would be giving me a lesson as he knocked up a frequent fifty with his consistent skill at long and short ginnies in off shots.

After dinner, I would often stroll on a clear, starry, tropical evening as far as the Gateway of India and have a chat with someone who was enjoying the harbour activities. It is always rewarding to try to ascertain what are the things we have in common with people of other races. One thing everyone liked to do at the Gateway was to buy a young coconut for *do anna,* about two pence in our old money. The milky white fluid and the soft, pulpy flesh inside the shell slipped down like nectar of the gods, and with a spot of gin no one could ask for more.

Dengue fever was quite a common ailment in Bombay. I contracted it a couple of times after being careless over exposing my head in the sun. High temperature and heavy headache, as if one has been kicked by a mule, are the symptoms.

Back in my room at the Warners' home in Mody Mansions, their domestic servant who waited at table appeared to soothe my throbbing temples with his cool massage of lean brown hands. It seemed to be effective; somehow I related it to the laying on of hands in the Christian religion.

There came a day when we all sat round the table in David Warner's home, having our lunch, Mr. and Mrs. W—— and Bertha, their daughter, Thelma and Bobbie Green, who worked with an oil form, Caltex (India) Limited, all good Eurasian folk, two Englishmen, and me, a Northwalian. There was a shout from the street, one floor below us, and the servant with the Sikh headwear went to the window and announced that there was a Gurkha officer in the street.

Some of us went to the window, and there, peering over the hedge in front of the house, was a large bush hat atop a big handlebar moustache covering a round face. It was my university friend Neil Cecil Højgaard, whom I had not seen since June 1938, seven years previously, though we had exchanged a few letters in the interim.

"What on earth is that on your face?" I asked. "It must have taken you since last we met to grow it."

"You blighter," said Neil, "I only kept it for you to see it. And now it's coming off straightway." He went straight into the bathroom and emerged with the adornment under his nose more in keeping with his rank of captain in the Gurkha regiment, his five feet three inches of jungle green attire looking every inch the part he had carved out for himself, starting as a private (sapper) in the R.E.'s and earning his promotions the hard way.

I made one more move to new digs when Ballard spoke of a vacancy in his place. There were several people from the office there, and the company was welcome for the last few weeks I was to remain at Bombay. The son of the house was a lieutenant in the Royal Indian Navy, not a very mighty service altogether at that time.

VJ day came and my wife wrote for me to apply for repatriation, but the Admiralty would have none of it and I was to find myself posted to Singapore early in April 1946.

LST 3011 was to be my conveyance from the docks at Modi Bunder. These landing craft, the large type, were all welded jobs. En route, we put in at Goa, Portuguese protected, from whence Rodriguez our office driver, hailed, and quite a few of his countrymen went ashore there. I visited the Franciscan church and, if my memory serves me right, stood on the grave of Dom Vasco da Gama, buried vertically under the floor of the church.

Da Gama was the great Portuguese explorer, born in 1460, died on Christmas Eve, 1524, just before sunrise. He was Count of Vidi Gueira and succeeded Albuquerque as viceroy of Portuguese India in the last three months of his life.

After leaving Goa, we sailed on to Singapore, calling at Colombo but not going ashore, and I was sailing on to another episode in my life.

9. Tidying Up

Remember when I reported early one morning to the Liverpool office, as an A.C.E., it was to go straight into work after a wretched overnight journey from South Wales, with a perishing three-hour overnight stop in a siding at Crewe. Remember, also, when I reported early one morning to Ewart House at Bombay as a C.E. it was to go straight into work after a wearying long journey by train from Colombo up through Ceylon and across the subcontinent of India. It was also to take over a double promotion for the first week and hold the fort for the S.C.E. India North.

Well, my arrival at the naval dockyard at Singapore Island (Seletar) late in March of 1946, after a few days chugging away from Bombay on LST 3011, was apparently going to be nice and peaceful, the morning being hot and the sky clear as I stepped ashore to wait for my black officer's trunk to be offloaded onto the quay side by my big canvas hold-all. Then, sailors saw that the stuff was on the back of a plain, open, flat-bottomed truck to move me in to transit accommodation at the dock-yard. I got in the truck alongside my gear, and then all hell let loose. No, it wasn't a revolution by the Japanese P.O.W.'s. It was the heavens opening! For those who have always been domiciled in a temperate clime an electric storm in the tropics can be something that has to be seen to be believed. To cut the story short, I turned up to report to Higgin-botham, now on seat as S.C.E. Singapore, looking and feeling like a drowned rat.

To make matters worse, my shorts and shirts, so nicely laundered on the ship, were ruined by the canvas dye of the hold-all bag as effectively as if the whole lot had been plunged into a hot bath. The replacement gear from store was not exactly the most stylish cut, made for slim, athletic sailor boys, which I was hardly capable of showing off to its best advantage. There was not sufficient air movement up the trouser legs for my liking.

The works staff at Singapore included a number of C.E.'s and A.C.E.'s about 50 percent of them sporting their R.N.V.R. uniforms, the rest in civvie wear. There was a good meal in the mess that evening after R.W.H. had shown me how to lose a few more pounds (weight) playing him at squash, my introduction to the game. In the U.K., it is pretty exhausting. In Singapore that evening, I was almost down to a grease spot.

The following day I was allocated an Austin twelve-horsepower utilicon vehicle, a light runabout with canvas hood, and syce. The drivers, went with the allotted vehicles. They were mostly Malays from upcountry, and good drivers they were. We sped on our way from the dockyard area on the north side of the island southward down the Bukit Timah Road and turned up at the staff quarters for the outstations establishments in the Tanglin area, just outside Singapore town. I was to be C.E. (O.E.) under the wing of J.W. Hooper, S.C.E.(O.E.), who was responsible for the conduct of all Admiralty works on the Singapore Island and some adjacent islets, though aerodromes were excluded from Hooper's purview, these being under the wing of the S.C.E. (Dockyard) F.E.P., Clear.

It was interesting to note as we journeyed from the naval base to the town that when we stopped for petrol at a filling station at the side of the road, there were no formalities. We just went in, got filled up, and drove away. In the immediate reoccupation and rehabilitation period that was the form for naval personnel. Presumably, red tape is all cut, or rather not introduced, as such niceties as form filling and vouchers and so forth are given a low priority in relation to the essentials of the action to tidy up the mess.

After my luggage was conveyed into my largish bedroom of the Colonial type house known as One Tree Hill, the syce ran me over by early morning to meet my new immediate superior, James Winston Hooper, the superintending civil engineer in charge of the outstations establishments offices near Tanglin Park. At the time of my arrival the level of rehabilitation works was such as to keep the staff busy, to put it mildly. Former naval and other service establishments had been desecrated either before the Japanese came, as a matter of the scorched earth policy, or as a matter of their occupation.

The S.C.E.(O.E.) himself was taking a personal interest in major rush jobs, like the repair of a jetty at a place called Loyang, which had been breached; the work consisted of the construction of a length of causeway to span the breach between the two sound ends of the original structure with its piles and framed superstructure. Another work of urgency was the building of a new fleet canteen, where hundreds of Japanese P.O.W.'s were working wonders under the watchfully strict eyes of their own officers. To have seen the Japanese servicemen at work then is to understand the success of industrial Japan in the postwar period.

I was replacing Comdr. R.N.V.R. A. M. Lawson, C.E.(O.E.), one of the C.E.'s I had assisted when at the Liverpool Works Office, and Hooper also had C.E. Waddington as a divisional officer. A.C.E.'s were Lts. R.N.V.R. J. K. Bithell and R. Bainbridge, mature men buzzing around their territories in the outstations areas in their Austin 12 Utilicon and

Civil-engineer-in-charge Admiralty staff group at naval dockyard, Singapore, 1946. Author is second from left in first row.

American open-air twenty-four-horsepower Jeep, respectively. A.C.E. R. Bailey was also to assist me again, on transfer from Bombay. It was always like that in the naval service, bumping into people whom you knew, so you had immediate rapport with some of the people around you, setting you off on the right foot, so to speak.

John Bithell was a tall, good-humoured man of artistic taste who floated around his area, with work centred mostly on the Singapore town. We shall meet him in a little tale I shall relate. Bainbridge was of north of England stock, keeping "Johnny Jeep," his syce, hard at it as he took in the outstations farther afield on the island. He had been in Sydney before his posting to Singapore and took a dim view of the adverse comparison of the two postings. *Push poor* was his favourite euphemism for Singapore.

The offices and the quarters at One Tree Hill were quite close together, and some decades later I find it difficult to separate the goings on in the office from those in the house accommodating us senior types, the engineers, plus Robson and Davis, surveyor and assistant surveyor of lands, respectively.

In truth, there was a refreshing stimulation in this rehabilitation work, rebuilding instead of destroying, and the complementary use of free time. There were spare-time leisure activities like Sunday picnics at the former Japanese naval base, an unfinished folly at the north side of the island, midnight swims at the *pagar* at Johore over the causeway spanning the Malay Strait, and occasional Sunday trips farther north, as far as the reservoir, Burong Pulai, in Malaya, so vulnerable to the Japanese when they invaded via the Malay Peninsula from the north. Still the Japanese had had their day and now had to do as they were told until the British authorities saw fit to release them.

Talking about picnics, S.C.E. Hooper, a bachelor, used to enjoy our office do's out on a Sunday and, in his quiet way, was a force for stability in the efforts and morale of us expatriates.

There was a tennis court at One Tree Hill where enthusiasts like Davis, Bainbridge, Robson and I played in our swimming trunks, due regard being paid to the temperature and humidity. One day a couple of bods started digging in the corner of the tennis court as four of us carried on playing. They turned out to be people from the War Graves Commission and exhumed two skeletons of soldiers some four feet down. Seeing men there, they were no longer a statistic; it was a tragic sight. There are other personal stories of those known to me whose memory I honour as they lie in Singapore, having perished in the service of their country. I will refer to these later.

One thing I have just remembered. the C.E. I was relieving, A. M. Lawson, formerly of Liverpool Office, was as dour a Scot as you would

Eastern shots. (*Top*) Bombay in 1945—the Taj Hotel, the Gateway of India. Singapore in 1946: (*middle left*) "One Three Hill," Tanglin; (*middle right*) the Johore Causeway; (*center*) Changi Gaol, where Japanese POWs awaited trial; (*bottom*) one of the "Worlds" of entertainment.

find anywhere, but when he learned of my plight with my lost attire, including two pairs of shoes, he was delighted to find a disused pair of his own fitted me perfectly, so when he departed for the U.K. I quite literally took over his shoes!

I promised you a story in which Bithell and I played our part, and so I am writing it down as I wrote it a while ago, a tale largely factual, though officers' names escape my memory. Let us call the story "A Slip-up in Singapore."

A Slip-up in Singapore

The dark grey cloud was getting bigger. The bigger it got, the better I liked it.

Seated at my desk in the Headquarters for Outstations Naval Works near Tanglin Park, Singapore town, I kept glancing to my left through the opened louvred shutters of the window, the movement of my head punctuating the routine of attending to papers mostly tabbed "Action this day."

Across the harbour some two to three miles off shore to the south lay Pulau Blakang Mati, the largest of various isles and islets to the south of Singapore's main island. Prewar accommodating about three hundred personnel for the British army, it was now home for some twelve hundred naval types in larger barracks. Apparently the Japanese had not used the barracks during their unwelcome stay and had left a full reservoir over there, but the water level was now reportedly dropping appreciably.

The heavy tropical rain precipitated upon the town just as I saw the bottom of my in tray, initialled a circular, and tossed it over my empty desk into the out tray, happily full. Stretching my back over the chair with fingers locked behind my head, I breathed in the cooler air from the now closed louvred shutters. To all intents and purposes I was up to date.

I opened my ears. The messenger's footsteps up and down the stairway of the old Colonial building, the chatter in the drawing office, and the thin whistle of an individual happily detailing a drawing now registered harmoniously on ears previously insulated from such sounds, of no consequence to the head to which they were attached.

It was two weeks since I had arrived at Singapore Naval Base from Bombay on LST (landing ship tank) 3011. The base was on the north side of the island near the Johore causeway, which gives access across the Johore Strait from Singapore's main island to Malaya. As I was new divisional civil engineer in charge of outstations works, my time had flown in absorbing so many files, papers, and affidavits on people perished in the fight to forestall the Japanese, sketchy reports on derelicted depots, and tours of inspection of their physical condition and the rehabilitation works in progress.

One outstanding item had been niggling in my subconscious. A.C.E. John Bithell had mentioned a couple of times, perhaps a little too casually, that he would be pleased to see the rainy season begin to replenish the said reservoir, now slightly low. He had been quick to reassure, however, in passing, "Nothing to worry about,"; "Very familiar with the locality,"; "Been expecting the rain any day now,"; "Was consulting water engineer before the war"; and "Well under control."

My first visit to Blakang Mati had not seemed to demand special urgency, and it happened that it was to be about the last place to be visited on my list, my only associated thought being of a half-full reservoir on a tropical island.

Lt. J. K. Bithell breezed into the office, as he usually did, a tall, dark, happy man with a local social life bringing personal charm into the execution of his naval duties.

"Taken a turn around the deck, Bithell?" I asked. His name was a point in his favour, that of my maternal Taid Bithell, master mariner in North Wales and boyhood hero of mine. "No problems at Blakang Mati?"

He hesitated, shaking his head with a wry, dubious smile. "No joy yet. Clouds missing the island. No help at all. Can't be long now. Law of averages."

"Is it still under control?"

His response was a fraction slow in coming. "Still monitoring the water level," he said noncommittally. Did I detect a shadow across his face?

The early morrow found us both departing, complete with twenty-four-horsepower jeep and Malay syce on an LCT (landing craft transport), from Ballard Pier heading across the harbour to the island in question. We sat in the transport as the syce took the wheel and drove off the landing craft ramp onto the slipway and inshore up a narrow paved road winding its way through the thick, green tropical vegetation of trees and undergrowth at either side.

At an elevation of some one hundred feet above sea level, the route wound to the left to open up a vista of clear, blue sky. A long, narrow valley lay ahead on our right side, the opposite side to the road being as steep as the Rock of Gibraltar, but green, despite the end of the dry season.

"Nice-looking valley," I commented. "How far on to the reservoir?"

"That's it," said my assistant.

"But there's no water at all."

"Ah, not to worry, just the top end here. Pond no. 2. Backs up from that dam."

He indicated an earth embankment across the valley in the distance, just discernable after his due warning. It was disguised by greenery, now within three hundred yards.

"There's a balancing pipe through the dam, for control." He was fond of the word *control*.

Beyond the dam, the road again wound to the left, as did the valley but farther ahead there was a right incline, taking both out of view.

"So this is the lower pond," I said, a bit lost for words, let alone the sight of water.

He nodded. "Water's further on. Soon be in sight."

The reservoir access road ended with the pulsating sound of a heavy engine driving a ram-type pump. We vaulted out of the Jeep to survey the scene. The main dam was an earth embankment spanning across the valley to a height of perhaps twenty-five feet, the upstream face bare and brown. A dark green, slimy algaic growth covered the valley floor adjacent to the dam like a thick carpet fitting from bank to bank.

I threw a cobblestone and saw mud plop through the puncture pile. through this mire stood the outlet pipe, its purpose to decant water from the reservoir when normally holding water. A solitary vertical finger of pipe was twenty feet high above the dirty bottom, a bellmouth at the top and four simple gauze-covered inlets were at the sides for circumstances when the water was getting low, and now only the bottom inlet allowed the filth to slurp through into the decanting pipe, allowing the liquid to go on its way to a simple system of sand filter and chemical dosage treatment into a sump of cloudy water having a smell of its own. As a boy I had seen better-looking liquid pour from a new sewage works my dad had constructed at Congleton into the river Dane.

"Ah, but there'll be some water stored in the subsoil," consoled my soul of support.

I had no answer for my friend's optimism. To see the high-level service reservoir or tank we struggled and slipped up earth-packed log steps alongside the cast-iron pumping main, its leaking joints squirting water with every pulse of the pump far below us. When we looked in to the tank at the top, it appeared that it had a capacity of about fifty thousand gallons, but now the flow was across the floor, no storage buildup in the tank was developing, as the barracks were draining every drop.

Still on the island, we made our call on Capt. Henry Bliss, who received us warmly as we entered his first-floor office. The barrack square lay to our right through the window and to his left.

"Tot of tiffin, gentlemen?" he opened.

After a bevy of banalities I sipped my tea, put it down carefully on his desk, and came briefly to the main point. (This old salt was still at sea and had to be brought ashore.)

"Ever been to Gib, Captain?"

"That takes me back. Wish I were there now," he reflected.

"How much water were you allowed?"

"The one bad thing, really. Two gallons a day, I think. Not a lot per man. Still, we managed. We don't waste fresh water in the navy, what?"

Standing together, we looked out of the window as I painted the picture of imminent water failure, the perfidious cloudbanks, the line leakages, the abnormal complement of island personnel, and last but not least, the effect of what was going on before our very eyes, the sluicing down of the square and swabbing it over to simulate the quarterdeck of a ship at sea.

"A hundred gallons per head of fresh water is being lost right there, Captain."

My flat statement struck home. "My own b—— fault," said Bliss. "Thought I was still at sea. No more swabbing the decks, what?"

"That'll help," I agreed, still sick in my stomach. "We'll have to see what else can be done." I was thinking with dread of typhoid, cholera, and amoebic dysentery.

When we arrived back at the office our phones were burning. W. H. Hammer Ltd., the suppliers of water to ships, were soon filling a three-hundred-ton lighter at Empire Dock, near Ballard Pier. At the same time, R.N. Base at Johore Strait was loading up iron pipes and flexible Victualic joints. Personnel with plumbing prowess were alerted for work.

The next day, the stricken island was being relieved as a mile or more of water main was laid on the surface from the landing hard up the torturous road as far as possible and then at a beeline for the tank at the top of the hill through the woods. The lighter's pump merrily charged the newly laid pipeline with *aqua pura* filling the storage tank and eliminating the need to use the dregs of the reservoir.

Four days later, the operations captain was chairing the flag officer's ops meeting at naval H.Q. near Ballard Pier. Bliss and I sat opposite him, with other members all round the broad expanse of table. When it came to me, I formally reported the need to import water to Blakang Mati, although the chairman had been kept au fait with the situation all along.

"A case of 'in ignorance was Bliss,' " quipped the oracle, looking over his half-round glasses with mock severity across the table.

There was no answer to that, but on our side we could see the midmorning sunlight framing a dark cloud as it passed across the window to leave a clear blue sky behind him.

That afternoon there was a torrential rainstorm, and that time it did fall on the unforgettable island of Blakang Mati.

At the end of chapter 5 I mentioned Leigh Hunt, the officer in charge of works at Milford Haven, who introduced me into the Admiralty in the first place.

Here, in Singapore, I was going through papers of the last days before the fall of Singapore Island and Leigh Hunt's name cropped up again. It appeared that he was serving in Singapore and was ensuring personally that the scorched earth policy was carried out. With the imminent arrival of the invading army, so read the affidavit, he was last seen turning the valves to empty the tanks as the flames shot high at the Normanton Oil Storage Depot. He was the sort of man who would have seen his action as just doing his job. He was a patriot I remember with gratitude.

Two other servicemen I was proud to track down, sadly in their graves, as they had passed on in service to their country, were Tom Ellis, a soldier from my boyhood village of Woodlane and, our next-door neighbour in Pembrey, John Mansel Lewis, a young flying officer who went down in the Malay Strait. Both these boys I knew to be so full of life before the war, and it was a privilege to stand before each of their places of rest and place a wreath in respect to their memory, as I was possibly the only person on the island who remembered them. The War Graves Commission did a tremendous job establishing the cemeteries where the dead were with their comrades. What can one say?

I was glad when a young man from my former home village, Ewloe Green, turned up soon afterwards. He was Ernie Peers and had had my address from my father. Ernie was in on a merchant navy ship and joined us all at the mess at One Tree Hill for dinner. The conversation took on a new dimension with a new outlook on life, we all having virtually exhausted our repertoires during the months we had been living in the same house.

One day the S.C.E. called me in to his office.

"There's been an accident in Java," he said. "An oil jetty in Batavia, struck by a Corvette. You'll have to go over and sort things out on our works side."

I could go by air or by sea. Having absorbed the picture in perspective, I decided to go by sea. After all, the incident had happened. There had been no loss of life. The damaged jetty had already been brought back into use by the Dutch authorities. All I had to do was assess the amount of damage and confirm this with the Port Authority at Batavia. Important, too, was the prospect of two and a half days trip at sea.

It was therefore another LST that carried me to Batavia to see the jetty damage, LST 3508, under the Command of Lt. Cmdr. R.A.N.V.R. G. M. Dixon, D.S.C. on a pleasant two and a half day's cruise to the Indonesian island of Java. These ships were pleasant to sail in, just small vessels, perhaps two to three thousand tons. The officers' saloon was small but comfortable, and I enjoyed the company of the crew immensely.

It is memorable to me that I enquired of one of the lieutenants—was it Smith, Stringer, or Urquhart—who had spent time in Singapore and in Java, "How are prices in Java?"

He thought carefully, then said, "Well, in my considered opinion, by and large you can buy for a Dutch gulden just about what you can buy in Singapore for a Malay Straits dollar." I had called on help from my Taid Bithell, above me, once again, as I had previously ascertained on the sea trip from Bombay to Singapore that you could buy for a Malay Straits dollar in Singapore what you could buy for a rupee in Bombay. On top of this I had found out for myself that a rupee in India was worth about the same as a shilling in the U.K. Now I carried with me a copy of the War Department Schedule of prices issued in 1939, seven years previously, and in the U.K. prices had inflated by 250 percent.

So now I had the means of assessing the price of any repairs to the jetty viz: WD price x Rupee x Malay dollar x Gulden all multiplied by 2.5. This was to be crucial to my mission.

Let us put this matter aside and step ashore in Tanjong Priok Harbour, Batavia, having espied the forlorn state of the oil jetty as we came inshore that August 1946. The end of the jetty had a section of quadruple-strength Bailey bridge cantilevering out to the point where the old concrete jetty deck must have reached before it was unceremoniously pranged by the Corvette. Two isolated bollards in the water were intended for anchoring the ships while taking fuel aboard, but "The corvette missed the bollards and ran its prow into the end of the jetty, taking an end section with it and fracturing the pipelines, with a consequent loss of a vast volume of oil into the harbour waters," according to one of the crew aboard.

When, therefore, I was picked up at the quayside and taken to see Captain Jones, I had had a glimmer of a picture of how the accident had occurred, but the hoary details were spelled out as he explained all that had happened in full, embellishing the story from the record of the inquiry he had held so that he did not miss a thing. It was then that I learned of the extent of the vast loss of oil sustained as a result of the happening.

"How can I help you further?" asked Captain Jones, R.N.

"I would like the use of a diver," I replied.

It was now late afternoon. "I'll have one down at the jetty in the morning," he promised.

The next morning a car took me over from the quarters allotted to me by Jones to the dock gates of the harbour in the control of the Dutch East Indies authority, the *Haven Direktor*. Two hundred yards away was a railway track from which snipers were having pot shots at the dock gates every now and again. On the way to the docks, the driver of my car had drawn my attention to the dismembered bodies floating in the canal. There was unrest in Indonesia at the time, and the Netherlanders were not exactly popular with the indigenous population. Fresh Australian food was not available to the few Britishers at the port, as the Aussies were not prepared to deliver; why I do not know. We were eating canned food, and I'll tell you about that later.

I made my way to the jetty, and while the diver was making his underwater survey I recorded my measurements and made sketches of the overwater situation. By the end of the day I had enough from the diver to confirm what appeared to have been knocked into the drink, and, with a dozen photos to illustrate my ruling dimensions I felt confident about my grasp of the extent of the physical damage.

The next morning saw me in the office of the *Haven Direktor*, checking the cost of the damage with the *Haven Ingenieur*. The latter worked out the cost of replacing the reinforced concrete at, "fünf hundert guilden per cubic metre." I worked it out from my preconceived notion of 1939 War Department Schedule rates, converted up to date and into Dutch guilden. We agreed within 4 percent. We split the difference and that was that. Now I raised, somewhat tongue in cheek, the question of "betterment," and the engineer came straight back at me with the valid argument that a new patch in the old jetty would not be an extra asset to them—no benefit at all.

In the circumstances, I was inclined to agree for the sake of international relations as well as the sense of the matter, but just then in walked the *Haven Direktor*, who was not so well versed in the King's English as his *Ingenieur*.

"How you are making?" he asked.

I said, "We are agreed, but you will be better off with a new section of deck, some x pounds."

"You have agreed it is better, fine," he said, and out he went.

"Well, if he says so," said the *Ingenieur*, and we agreed on a betterment allowance amicably.

Back in the Singapore office, Sweetinburgh, who ran the drawing office, was pleased with my data and soon produced a beautiful blueprint drawing to illustrate my report to Admiralty in the U.K.

Healthwise, I was not so pleased. Out of the trip I had had the disbenefits of a filthy cough and food poisoning from all the tinned food, and it was two weeks before the naval medic managed to stabilise my condition, while I continued to work with a sore behind.

The Japanese may have left Blakang Mati alone, but they were busy little men at Normanton Oil Depot. There were underground workings and high-explosive "gelly" was oozing out from boxes as Bainbridge, and I went into the tunnel to make an inspection. Rubber boots were a must, as a spark from a boot nail could have set off one almighty explosion to rival the one at Bombay docks just before I had arrived at that place.

It was a matter of some relief to us in the civil engineer in chief's department when the navy explosives personnel came in and removed the stuff. In such a place one is grateful for the expert.

On a Sunday morning jaunt to Pulau Bukom on a D.U.K.W. (an amphibious tanklike vehicle) the S.C.E., James Hooper, was taking us for a picnic. We were looking forward to a leisurely day of pleasure lying on the beach, enjoying light refreshments. It was to turn out rather differently. The ponderous little craft chugged over from the main island of Singapore, making for the islet of Pulau Bukom in the choppy water, the rest of our party of chaps from the office hanging on, owing to the singularly awkward motion of the type of vessel, when all of a sudden the engine conked out.

The D.U.K.W. had a huge flat bonnet covering the engine, really heavy, with a metal strip all around the edge to form a lip for the rain to shed clear of the engine. A few of us lifted up the bonnet and peered at the engine, hoping to see a stay to prop up the bonnet, but there was no such facility, nothing at all like that.

Then someone fished out a brush and stale that had been stowed away in the craft, apparently for brushing down, keeping the craft shipshape and navy fashion. So the stale was used to act as a stay by propping up one corner of the bonnet, sticking under the right hand corner as we looked at the engine. Then we were all peering under the bonnet again, five of us fiddling at once to see if we could get the infernal machine to restart, while Hooper stood back puffing his pipe, enjoying the silence of the water lapping at the stranded boat. Presently, when it was obvious we were not having any success, the S.C.E. came over to see for himself. The commander rested his right hand on the brush stale and leaned in towards the engine, his weight carrying the light prop forward, and down came the bonnet with its lip acting like a guillotine, cutting into his cranium.

We all automatically took our heads from under the bonnet; my own prying pate, having been under the hinge end; would have been completely severed had not Hooper's skull caught the decending might of the metal member. With one accord we held the weight of the bonnet and put it down on its seating.

James Winston Hooper stood puffing his pipe with the determination of Popeye.

We looked in amazement. "Don't you feel anything?" we asked.

"Should I?" asked the chief, looking the colour of chalk as a trickle of blood ran down, thick and red on either side of his face from the weal on top of his cranium like something a butcher had inflicted on an animal.

We were becalmed, stranded, defunked, and with a patient on board. He had been felled like an ox, yet stood up as erect as a figure in Madame Tussaud's.

Just then there was the sound of a motorboat putt-putting towards us to see if we needed any help. It was a dinghy with an outboard motor, and Hooper and two of our members were off on their way to the north side of Singapore Island from where they had to make a tortuous journey down the island as far as Ballard Pier and over to the sick bay on Blakang Mati. For our part, we eventually got the D.U.K.W. in tow and returned to base on a downbeat, but I remember the upright figure was at work after receiving his stitches and a plaster and carried on with his duties of the week without any complaint.

The time was soon to come when, with work winding down at Singapore Island, old Hooper would precede me to the U.K. and I would be serving as officer in charge of works there, for the last three months of my service out east.

Casualties that meant losing colleagues in an undesirable way occurred from time to time, and I can recall three such occasions. One of our supervising staff in the field, a foreman of works, met a fatal accident coming home from a social evening. It is a very sobering experience to consider how best to write to tell a person that her husband is dead, and one does not ever get used to the responsibility for breaking the news in an appropriately sensitive manner.

On another occasion it fell to me to escort to his boat anchored off Ballard Pier one of our number whose rationality had been affected by the sun. It is sad to see a man's attitude change as a result of a mental breakdown.

126

Then it came about that Roger Higginbotham, the S.C.E. himself fell ill and was committed to hospital with some complaint that laid him low for a longish illness. I used to pay him regular visits, which he always appreciated. Later I was to meet him again at his home in the U.K., and I will touch upon this on another page. So although I was stationed at Singapore for only few months, I was to see at least half a dozen people close to me who either died or turned out worse for wear healthwise.

Let us try and finish on a lighter note at Singapore, a place that has long been a centre of commerce and has prospered particularly under Prime Minister Lee Kuan Yew since he took office in 1959. There was plenty of leisure activity available when I was there, for the country was in an ebullient, recuperative mood. People like Tommy Trinder, buzzing round in his personalised numbered Jeep TT1, performed at the Victoria Palace theatre and were to be seen at the Yacht Club or at the world famous Raffles Hotel named after Sir Stamford Raffles.

Then there were the worlds—the Happy World, the New World, and the Great World, all huge open fairground places where one could be entertained or eat in the open. Shows in square rings were intriguing. Some would be dancing to a tune with a hypnotic, repetitive rhythm, five beats to the bar.

It was at one of the Worlds that I witnessed an expert sword swallower, the sword act being followed by swallowing a pitcher of water full of fish, the gallon or so of contents disappearing down the gullet into the stomach, only to be regurgitated into the pitcher again, the fish swimming all the while.

Cabaret rooms were available in the Worlds where one could buy a drink or have a spot of food, where "Chinese" never tasted so good, or again, one could buy a supply of tickets and jig around with the hostesses, slim, young girls in their typical short dresses, slit at either side and known as *cheong sams*. I suppose in the swinging atmosphere of the aftermath of the war there was very little one could not have in that vast free port.

At the end of 1946, word came through that I was to be repatriated, and eventually the *Britannic* set sail from Singapore, under the command of its master, R. B. G. Woollatt, R.N.R., with me aboard, having said my good-byes on the nineteenth of November. The *Britannic* was a big ship of twenty-seven-thousand gross tons, a member of the Cunard White Star Line, registered in 1930 at Liverpool, where else? Acting as a service ship, it was returning three and a half thousand passengers mostly, navy, army, and airmen, and carried a compliment of 535 crew. Our route lay via Colombo and Bombay and across the Indian Ocean through the Red Sea and Suez Canal. By the twenty-ninth of November we were leaving

Bombay and sailing for home at a rate of about eighteen knots.

The Red Sea was infernally hot. In the after deck hundreds of air force lads were lolling about in the stinking heat on the hot decks. It seems there was much complaining and one young type could bear it no longer. "Why don't you jump overboard?" may well have been the suggestion. Anyway, the rumour was that he said, "I think I will jump over the side," and over he went, into the boiling hot shark-infested sea. The alarm being called, the first thing most people knew was when the ship's siren sounded a number of warning blasts and an accident boat, no. 2, was lowered into the water on the port side, the big ship turning in the water in the hope of effecting a pickup. Ship to ship communication must have been immediate, as a French ship also in the vicinity joined in the search. When finally, after about half an hour, the Frenchman found the boy's cap floating in the water, it was known the sharks had had their fill. There was continued searching, but in the end it had to be abandoned. "It was at 3:20 P.M., the 5th Dec when LAC W. Hickey, RAF, a service hospital patient, fell overboard from A deck, port side, forward." So recorded the ship's log. "No trace of the body was found."

There was depression on the ship to have lost a young man on his way home. It was a topic of conversation we could have done without.

Putting in to Suez, we purchased leather bags and pouffes from salesmen on boats down below. Soldiers a couple of decks below us were haggling fiercely about price with the befezzed vendors below them in the water. For quite a while the bargaining went on cheerfully. The reddish leather goods, mostly bags, were coming up in a steady stream, and money was going down in payment. Then one soldier absconded with a bag without paying the money. This upset the particular vendor, as he waited for the money before being prepared to send up any more bags. The temper of the sales was immediately altered, abuse now being rained on the English by the offended waresmen, with good reason. "Eenglees gentlemen, Eenglees peegs," was a favourite epithet, and I must say the whole happy atmosphere of the trading had been spoilt by the wild boy who had made off with a free article.

One final act of treachery happened to me as the ship lay alongside the Princes Landing Stage at Liverpool, ready for our disembarkation. My suitcase lay on the bunk of my cabin while I attended to the call of nature. The last thing I had packed was my dressing gown, a present from the staff at the Liver Building before I had departed for the Far East.

When I got in my train at Lime Street station to travel down to South Wales, I opened my suitcase only to discover that some light-fingered type had whipped my dressing gown—just like that! C'est la vie!

It was December 15, 1946, when I made my way home after service overseas, and I now had a few weeks' leave before I had to resume duty in February 1947. My run up to starting work in the U.K. again will be described as the commencement of my next chapter. Cheers! . . .

10. Casting Off

For much of the five weeks' leave spent at Pembrey, Lin and I were looking for somewhere to live near Pinner, Middlesex as we were given to believe that that was where I was to be posted on completion of my leave. Pinner was then the location of the Admiralty civil engineer in chief's headquarters. We searched all over for a house but were not having much success, such properties as were available to rent or buy being beyond our means. It was a time when rebuilding had not yet begun to make any impression for prospective buyers.

We need not have bothered, because the Admiralty changed its mind and when it was time to return to duty I was posted to Rosyth Dockyard. I remember thinking what a thing to do to a man just returned to the U.K. from the tropics in the winter, and anyone who remembers the 1947 winter will know what I mean. Luggage in hand, I left the hot coal fire in the Cliff, said farewell to my family, and went off to travel to Scotland. The train journey took about twelve hours through the night, and there was the stopper on through Inverkeithing to Dunfermline, where I booked in straightaway at an hotel. The journey had been long and tedious, yet it had been refreshing to talk to the people I met on the train. At such a time one's destination assumes living identity, no longer being just a place on a map, and one is charged with anticipation.

My bags in my room, I rang the works offices at the dockyard, and they promised to collect me after lunch, so, with an hour to spare I had a bath, then walked round the town to stretch my legs. Toddling round the town that morning gave me a quick impression of the local geography. The Andrew Carnegie legend was what first implanted itself into my mind. That philanthropist, a self-made man and oil magnate, had set up libraries everywhere. The glen was a veritable wonderland of nature.

The driver of the office car made his call after I had my simple repast, and away I went with him to report to the C.E. in Chief's Department offices at the dockyard. A. E. Chatterton, S.C.E. was just on his way out, a tall, dark-haired, fresh-faced man in his forties with muffler and grey woolen coat to keep out the cold.

"Ah, Price," he said, "nice to see you. Good trip?"

I did not complain.

"I'm just going over to Dalmeny for the regular progress meeting," said this real Irish gentleman. "Why not come with me?"

I went.

Over in the car he gave me a quick appraisal of the works in the Rosyth area as we sped down to North Queensferry on the banks of the river Forth and in sight of the unique dominant Forth Bridge edifice spanning the waters to carry the railway across.

"Your job will be C.E. (Outstations)," the S.C.E. informed me.

Inevitable, I thought. Outstations at Liverpool, divisional C.E. in Bombay, with so much of the work in outlaying areas, C.E. (Outstations Establishments) for Singapore Island, and now C.E. (Outstations) at Rosyth. Chatterton had another divisional civil engineer covering the dockyard itself, by the name of Farrington.

We went down the stone approach ramp alongside a stone jetty, boarded the car ferry, crossed the Forth, and went over to South Queensferry, the car making a short run to the gates of a large new stores depot in course of construction and well in hand progresswise. Government-type corrugated roof and brick-walled storage sheds, loading platforms, and large sliding doors hung on Coburn or Henderson tracks to give access to the sheds were in evidence, and a large vehicle yard gave good manoevrability for the ship support transport vehicles. The establishment was already partly in use.

A.E.C. chaired the meeting with practised perfection. It was really a coordinating meeting between various heads of service department involved at Dalmeny, officers in charge of stores, electrical, mechanical, and other specialties. No question arising was insuperable, and I had the impression that had there been a major upheaval to contend with, it would have been absorbed with relish by the chairman, who emanated the confidence of one who had been through plenty during, and had emerged with credit from the recent worldwide conflict.

There were other naval works at the South Queensferry side of the river, which we dropped in to see on the way back, and eventually, after a short hour at my office desk to ponder over a few key files with an outstation flavour, I went back to my hotel room to contemplate my new life at Rosyth. The room was not warm, so I went to the public room, and as the physical heat seemed to lack the good cheer of the people, I took a walk out in the lighted streets to acclimatise my body to the fresh nip of the Scottish air. The smell of chips was compulsively inviting, and I soon had my hands warming in the package wrapped up in paper, opening them and indulging myself in the street like an eleven-year-old.

Back in the hotel, I wrote a few words home to Lin, her Mam, and the boys and, after dropping the letter in a box, went to bed. The first day was over. It had been a quick introduction, just as I had had to Liverpool, Bombay, and Singapore. It was just as well because although I did not realise it at the time, I was destined not to stay very long at

Rosyth, though I must say my first impressions had been of a very friendly people in an agreeable works area.

I had mentioned in my letter home that I could just about keep my head above water financially, as the Admiralty paid subsistence for a time while I was maintaining "two homes"—I didn't actually own one! It was just as well Admiralty rules made this possible, as my efforts to find somewhere for the family to live in the first month or so were so abortive. If only there had been a small building like a private garage we could somehow have managed to make a home, but even the tiniest crofts with one apartment up and down, were not to be found.

In the first few weeks at work my trusty driver, Mutter, conveyed me around the sites, places near the dockyard and also as far as the northern extremes of Fifeshire to Perth, where we had five depots. The winter was severe, with much hard frost and snow, and in this time of power cuts and shortage of food the climatic conditions were a test of British character all round.

At first the snow was light and cars and trucks blazed their trails through the shire roads, but later things bogged down and the road-clearing machinery of the local authority was overtaxed. Chatterton stepped in, in response to appeal from the highway authority, and naval bulldozers and angledozers stepped into the breach to tip the scales against the natural catastrophe.

A large shed, a Calender hangar, collapsed under snow load. Designing for normal snow load of perhaps a foot in thickness is one thing, but in the teasing situation of snow falling, sun melting it into water and overnight freezing, turning it into ice, and the cycle repeating time and again, a build up of solid ice had occurred, and the point of overload and collapse arrived.

Up came the civil engineer in chief in person to assess the damage. Sir Arthur Whitaker brought up his brother for company and an opinion. He could not have brought a better man. Now it occurs to me that both of these men had a simple, direct approach to life. When I went to Liverpool University to study civil engineering, the heavy-jawed, deliberately spoken professor of structures made the solution of networks seem like child's play and Sum "A" line approach was like poetry. If I can compare Professor Whitaker's mastery of his subject (structures) with anyone, it would be Professor Hulme (maths) when teaching Fourrier Series and such. Simple and effective, Whitaker was the sort of man who came along with the students to summer camp at Llangollen burning the porridge with the rest of us or sitting in his tent reading paperback westerns. The man's mind was fresh, lucid, and explicit, and in asking him for a second opinion Sir Arthur knew he had a reliable man.

Now what of Sir Arthur Whitaker? The story went of how, when he joined the C.E. in Chief's department he straightaway studied the family tree of the staff to see the best route for him to rise from the bottom to the top of the tree. Well, if the story was true, it worked, for there he was as the C.E. in Chief himself.

After about five weeks of duty at Rosyth, my family still living in Pembrey, Mutter was driving me back from Perth through the white countryside of Fife. Arriving in the yard of S.C.E.'s offices, we were met by colleague McLeod, an upstanding established-grade A.C.E. who told me there was a telegram for me in the office.

It turned out that Lin had decided to join me and ask me to meet her at Edinburgh station. What good news; she was coming from South Wales, having the loyalty and trust that I would be able to meet her with no time specified in the telegram! My driver needed no persuasion to give of his time and take me over to Edinburgh to try to meet the family. He was aware of a train coming up from the south late in the evening and was confident they would be on it.

As we kept a vigil on the cold station platform our feet began to freeze solid till finally in came a long, long train with two engines. Out onto the platform poured a madding throng. Three forlorn people from Wales were looking about, a young mother with plaid coat and woolly hat and mittens, and two little boys in duffle coats and sporting the red boxing gloves they had had for Christmas not so long ago.

It was a moving moment, a miraculous meeting, and I thanked my driver, Mutter, for his perception in coming to our rescue. On our way back to Dunfermline the penny dropped that my hotel would not have room for the family, my bedroom being far too small and the other rooms being full. Mutter was again able to suggest we try a private hotel, and it was to Comley House that we made our way, to find with relief that there was "room at the inn." So we spent the night at this haven in a large family size bedroom, with double bed and small ones for the boys. There was no fire in the grate as there was a dreadful shortage of fuel in those days.

The evening meal was frugal. After finishing it, we did what we were to do for most nights when we were at Dunfermline in those dreary postwar days—we bought fish and chips. Returning, we folded the paper tightly into cracker shapes and coaxed a small fire to warm the grate before we got into bed. It was my thirtieth birthday, February 6, 1947. I know because our second and now only daughter was born in Bridgend on November 13, 1947. (That will be part of the next chapter.)

From the moment Lin arrived with the rest of the family, so advised the D.H.C.O., Mr. Tonkin, next day, the regulations stated that my subsistence allowance would cease. So now there were hotel bills to meet for the whole family, plus fish and chip extras. This outlay alone was far in excess of my salary. Being together was, of course, of far greater immediate importance than financial considerations. We could regard the time in Scotland as a big vacation but, being something of a realist, I soon had a niggle in my bones that before very long I would have to overcome the Mr. Macawber situation of expenditure of twenty shillings and sixpence spent for every twenty shillings earned. Nowadays we call it *adverse cash flow!*

Meantime, on with the life in the frozen north. We did not stay long at Comley House, as it closed as a private hotel and was put up for sale as a private residence, which it had probably been in years gone by. It was a nice, big, classical house with square, solid architectural lines.

We were fortunate in finding another family room at the Brucefield Hotel. The boys were attending a primary school across the park, near the Glen and used to return home trudging through the snow in their rubber boots, their healthy cheeks belying the empty-tummy situation that was inevitable at the time. The change of schools they had had—this was their third—had not been conducive to the continuity of their education. Pembrey in South Wales, Wirral in Cheshire, Pembrey again, and now Dunfermline in Scotland, with different technniques of teaching, particularly sums and different accents, were enough to unsettle even the brightest child.

One day Lin was meeting the boys from school when she encountered the schoolteacher, who spoke to her in strong Gaelic accent: "Are yewr boys deef?"

"No, I doant think so," said Lin, herself of longstanding Welsh-English accomplishment.

"Well, Ah dinna theenk thay kin heerr me. I hidda poot them in the frint," explained the good lady.

"Well," replied my patriotic Welsh spouse, "I doant suppose they can understand your accent. I can hardly understand you myself."

If my phonetic representation of the verbal exchange is poor, it only serves to highlight the difficulty of communication in a strange part of the country and helps one to imagine the dire consequences on a child's progress in school. This was another argument in my subconscious for having to return to South Wales, even if it meant leaving the Admiralty. If I was, to say the least, marking time in my career, my children were, if anything, going into reverse.

The final argument to motivate me to leave the service was soon to come from developments at work. Circulars were making it clear that a state of staff redundancy had arisen and that complements were to be wound down with the advent of peacetime conditions. New works were now going to be a thing of the past. So far as my wartime Admiralty experience was concerned, movements had always been forward-looking. Essential schemes for naval works, provided they were realistic, had seldom been delayed, and one developed a sense of timing. We "temporary bastard" types were ideal for running the gauntlet and anticipating approval of proposals and ordering work as emergency work more often than not. It was what won the war!

Now that the "permanent bastard" types were resuming their authority, especially in higher echelons, they were again able to cool the enthusiasm of work zealots like myself. A scheme for the reinstatement of a jetty I had under way, "in anticipation of approval," was frozen, and with it came my realisation that we were now moving into the "care and maintenance" era. My correct procedure would have been to apply through the proper channels for redundancy, but this was unlikely to have been granted, as with my record it was on the cards that I would have been offered a permanency, but my established grade would have been as A.C.E. in accordance with the standard procedure, with posting elsewhere and a drop in pay. On the other hand, if I had been granted redundancy I would have been given the princely gratuity of seventy three pounds and had to return to South Wales with no job. It was a catch-22 situation. I plumped for applying for a post with other people in the South Wales area before giving in my formal notice.

This was going to take time, so my subsequent time working at Rosyth and living at Dunfermline was spent under the shadow of hopefully imminent departure.

The people of Brucefield Hotel were sociable, friendly, and interesting. The young family proprietors were understanding of a couple with two young rapscallions who needed some elbow room. They could do no more than anyone else to add bacon and eggs to the porridge we always had for breakfast. The absence of traditional breakfasts at that time meant that the main course was the inevitable bap, with tinned beans. Those who have had beans every day know how "same-ish" it can be.

The pleasant atmosphere made up for a lot. What about the two weeks when the Glasgow players, including the late actor Roddy McMillan, came to stay at our hotel? Mr. and Mrs. Jenkins, friends of ours, as he was an officer in the Admiralty Stores Department, came with us all from the hotel, and we watched Roddy give one of his early day performances of a play set in the Gorbals. The hotel was certainly agog when the players were with us.

Naval Commander and Mrs. Haggard and Procurator Fiscal and Mistress Henderson were two other couples staying at the hotel and gave continuity to our home life, such as it could only be for the rest of the time we lived in Scotland, in a period when any place was bound to be seen to a disadvantage. The legal gentleman and his lady soon moved out to assume residence as the new owners of Comley House, where we had first lived with them as fellow guests, but the Haggards and the Jenkinses were to remain after we had gone, and truth to tell, we have often remembered them kindly over the years, even if we have fallen out of touch.

A grand little party was held one night with our hosts the instigators. Friend Jenkins, an Englishman with a Welsh name and Gaelic connections, did fine performance of a Scottish Highland reel. In a turn on the piano I found myself in unusually tuneful accord with the company, causing an American visitor to pronounce what has remained my accolade of appreciation of my efforts on the piano: "You sure can tickle those ivories." Never in the life of this poor mortal has anyone before or since expressed a word of respect for my doubtful manner of playing the ivories by ear.

On the subject of social occasions, Lin and I attended a sendoff at the dockyard for a very senior naval officer, retiring at the end of his long service to the nation. It was a saddish, poignant occasion, though not intended as such. It is not the only time I have been present at the parting with one who has given all his working life to a single organisation. Those present are but a small sample of a vast pageant of those who have passed through in this lifetime. The occasion can be an anticlimax—the end of an era. As one of those wishing the gentleman a happy retirement, I felt somehow inadequate, unworthy to be among the invitees, not having been long enough to value his service in a personal way. Yet with hindsight, insofar as I still remember that moment of feeling, perhaps our attendance was at least the best that could be done by the organisers. Perhaps he saw not just the token number of recent colleagues present, but all those with whom he had served in his long career.

James William MacLeod was a good friend to me at Rosyth, a colleague nominally my A.C.E. but who worked to me with confident independence of action in the various works at the barracks at South Queensferry and other places near the river. Before my wife had joined me, MacLeod and his wife had shown me their kind hospitality at their home in Edinburgh and introduced me to the city. Mac had seen hardship as a P.O.W. under the Japanese in Hong Kong and lived to tell the tale, just one of those who, like Leigh Hunt, had been overrun by the Eastern invaders soon after Pearl Harbor had fallen and Britain followed America into war with Japan. His reward was to have all his possessions lost and a protracted correspondence with the Admiralty for some compensation, an experience adequate to give anyone a jaundiced view of the service.

Replies came at last to my applications for a change of job somewhere in the South Wales area. Consulting engineers W. S. Atkins and contractors George Wimpey and John Laing were interested. These resulted in three interviews in the London offices of these great firms. Eventually I decided to take up the offer of Geo. Wimpey to work as a sectional engineer on the construction of the mammoth new steelworks they were building at Margam, Port Talbot.

Well-wishers at Rosyth made us half-wish we were not going, but with the stringency of the times, especially affecting those in the north, many of our sympathisers were wishing they could be travelling south with us. The four months or less at Rosyth had been just as eventful for me and my family as many a longer period, and we left Edinburgh station for South Wales, each with nostalgic memories of bonny Scotland, albeit in a period of hardship.

Once again we were heading back to the Cliff, and all our hearts were lifting. Lin was going home. The boys would be seeing their old school friends, and I was casting off navy works to take a new direction, as Abraham Eyre Chatterton had put it, "to hustle work" on a big construction project.

It was the end of Rosyth and the end of the Admiralty for me as I left the service a thirty-year-old civil engineer who in the last eight years had been transferred four times and promoted three. I was not entitled to any gratuity, though I have always treasured the fine memories and generous C.E. in chief testimonial I retain from the Admiralty.

11. Hustling Steel

Margam was a mammoth project. It was May 1947. At Port Talbot, South Wales the old steelworks, black and smoking, was already being extended when I bumped into Irvine, the agent for Sir Robert McAlpine, one day on the train as I travelled from Pembrey to start my new job as a sectional engineer for George Wimpey and Company Limited, main contractors for the immense new works just starting, a works that was intended to lift the country's production of steel from 15 to 16 million tons per annum. From what I could understand from my fellow traveller, "Macs" had not the spare capacity to take on the extra work involved in designing and building the new works themselves and so confined themselves to the extension of the old steelworks.

Now W. S. (Bill) Atkins had the job of designing. The great man had personally interviewed me for a job in the same week as Mahoney of Wimpey had taken me on, on the construction works for which they had been appointed. I had earmarked myself a house in the area, but as soon as A. A. (Dick) Wright, Wimpey's agent, met me he pooh-poohed the idea and straightaway told me they would fix me up with a house themselves, on a rental basis.

It became fashionable in later years to have agents subservient to project managers, chief project managers, and area managers, as bureaucracy took hold of the expanding contractor organisations, but these types have so often been faceless individuals insofar as the contract conditions are concerned. Dick Wright struck one as being his own man, speaking for the contractor and responsible for undertakings he gave. Stand or fall, that was the way of the old school. It was clear the Margam steelworks was his delight and with immediate assistants like Bill Furlonger, subagents Chaplin, Lawrence, and Bedell, and general foremen for general site works, a strong and enthusiastic team was getting down to the scheme prepared by W. S. Atkins and partners, with Americans, the firm of I.C.C, the kiddies who had the expertise for the processing plant. Both these consultants were under the eye of the Steel Company of Wales, Chief Engineer Cartwright.

As I was sectional engineer, the setting out of the work, the time sheets allocations, and the measurement of the major east section came under my purview. Sixteen young engineers were occupied hard at it from eight in the morning till they finished any time after five at night, so often carrying on to complete their task.

"You're a bloody slave driver," said John Braithwaite one night as he returned at seven o'clock after getting the measure over half a mile wide and much longer length of the area of the site he was concerned with, the rainwater running out of the seat of his corduroy trousers. Then he laughed as I jested, "You look like a very leaky 'Braithwaite' tank."

Wimpey's effort on this job was first-class. The systematic brain of Dick Wright as he carefully listed the tasks of the day the first thing each morning on a clean sheet of paper on spotless blotting paper kept the wheels in motion for the rest of the work force on the site. It was apparent that as earthworks were followed by blinding concrete, by carpenters' formwork, by steel fixers placing reinforcement steel, and by concreting gangs filling up the forms, each gang would be on top of the gang in front by reason of being just that bit stronger in numbers, so the "hustling work" referred to by A. E. Chatterton assumed a reality for me. Bill Chaplin would come out with the news: "We're going for this next." It would be up to the engineers to set it all in motion. That was the way it was. That was the way it had to be. That was the way the impetus was maintained. Whether this was planned (and the firm was good at planning) or whether it was due to the instinctive experience of the general foremen like Peter and Harry is really immaterial to the resulting achievement of progress. I suspect it was a combination of both.

Atkins's staff were on site in strength, with an adequate design and supervisory complement. There is no doubt that they had a mammoth task in keeping the working drawings supplied to meet the inevitable alterations that emerge with experiences of continuing winds of change in all industrial development.

The Steel Company of Wales displayed a sharp sense of responsibility. Although the S.C.O.W. had delegated the works to responsible contractors and the schematics and detailed planning and designs to the two main consultants, one of whom was also mainly responsible for supervision, they retained direct discipline on setting out and finance with their own surveyors, both land surveyors and quantity surveyors, the latter being the firm of Geo. Corduroy and Partners.

The setting out tolerances were vital. When we, as contractors, wanted to know the accuracy tolerance for the giant six-inch-diameter, eighteen-foot-long holding-down bolts, the answer came via Atkins's staff: "No tolerance." Now this was a big site, and even the earth expands and contracts as heat gives way to cold in the daily round, but apart from that, it is usual to have a practical tolerance in all work. We were glad it was the S.C.O.W. that would have the final check before concreting in the bolts in question. Of course, alignment in a strip mill is absolutely vital and the templates positioning the great bolts had to be set to the very limit of meticulosity, which we at Wimpey's were capable of. But let us not stray from the general nature of a life story of this kind into the methods of construction of civil engineering works that are more properly dealt with and available in the volumes of the journals of the Institution of Civil Engineers.

When new works were being done in a big hurry during the war for emergency reasons, it might be opted to select a contractor with the necessary resources quickly, before all the taking off of the quantities or even all the drawings had been prepared. Civil engineering contracts were sometimes let on a "target price" basis. This offered a formula for evaluating a contractor's prices for all the various items that would crop up as the work was executed. The device enabled both the contractor and the employer to benefit from improved efficiency of the contractor, though it did all hinge on the initial "target prices" being fair in the first place. This was the sort of arrangement that got the work going at Margam, and the S.C.O.W.'s independent quantity surveyors served a vital role in their own work for the company. It was necessary for the contractors themselves to maintain numerous costing checks of the items of works on a vast scale and to keep up to date with allocating the gangs' efforts into categories appropriate for the items. I fell into the habit of arriving well before other chaps in the office so that the hundreds of time sheets for the men and the plant could be assigned with their correct symbols without my being interrupted by the calls of the day about to start.

The civil engineering construction, foundation works, earthworks, piling, reinforced concrete, and railways amounted to about £6 million of the total scheme, worth about £50 million (a billion or so in the early eighties), and to get out the bulk of this £6 million in about eighteen months was moving apace. On the east section alone, the largest of the three sections, we had a turnover of the then money of £250,000 per month, which again indicates that we were "hustling work," in the words of my old friend A.E.C.

Dick Losely came in as sectional engineer to help Bill Chaplin, and this helped a lot, as Dick had worked with Bill on previous jobs. My duties now included the coordination of work of the twelve engineers of the west section to ensure that the records were kept in a consistent form.

Quickly and surely the steelworks began to take shape. The original swampy site had moved through stages. There was desilting of the marshy ponds and back filling with sand excavated from the sand dunes or dredged by the Mears' dredger and pumped over the site. Over the whole flat area of the site there crept a two-foot-thick carpet of slag, which came from old works tips, blasted and excavated and transported and tipped into place. New slag supplemented this when it was tipped, hot and fresh, from the adjoining Port Talbot Steel works into a lagoon of water, the rapid shrinking of the molten slag causing it to shatter into millions of brick-sized lumps, ideal for winning by dragline, hauling and tipping and spreading as an oversite carpet for the vast new yard areas of the mammoth works.

At least twenty thousand piles were driven over the whole site by three specialist firms, Franki, West, and Simplex. Where acid ground attacked these they had to be dug out and replaced, using acid-resistant cement. Railways and concrete subways for services had to be laid and constructed. Heavy foundations to carry the steel stanchions of the mill buildings to house the great heavy machinery of the ingot stripping and stocking and the soaking pits buildings were provided as the development grew from end to end of the site of the new "hot process" works. The "cold process" was to be constructed later, at the end of the new works, remote from the town of Port Talbot. This would be done by a new contract to be let in the future.

My preoccupation with giving an outline of my working life spent at Margam from May 1947 to August 1948 has meant that I have so far not mentioned a word of what was going on domesticwise. The house my firm leased to me was at Bridgend, and my family duly took up occupation in July 1947, as soon as the children's school year at Pembrey was over, our two small sons being young and in infant classes. When he returned to the Pembrey village school after being in Scotland Newton, our elder boy, was indelicately told that he had lost ground and instead of being ahead in the lessons in class as formerly, he was now behind and "It was all the fault of him going to Scotland." This was hardly the boy's fault, but everyone knows how hurtful can be the odd clever remark of an overintelligent, half-wise teacher, and this is not to labour a point but merely to say that we all have our human failings in guiding young ones.

With a holiday period in front of them, the boys soon made friends with neighbours' children in Bridgend. They eventually started school in the autumn term, when they got on well with no complaints that I can remember. However, Bonfire Night was due in a couple of days' time when an incident happened to waken us up a little.

David came running in to the house, breathless. "Newton is on the fire," he declared, looking the colour of death.

I followed him out as he ran pell mell into an adjacent field. The field was empty save for an oak tree. There was a fire burning, and Newton was shouting blue murder. When I got to the tree I could see that he was tied tightly to the tree around the feet and the waist and chest, with his feet off the ground above a small fire of twigs that was beginning to flare up in the breeze, and I had to kick wildly to disperse it away from my son. It was some time before I managed to cut him free.

It seems that some bigger boys had done this for a lark while they were all at play! "They never meant it" was the sentiment of my boys, but I had to ponder long into the night before coming to the conclusion that they were probably correct. The boys' playmates had disappeared into the bad light on seeing me coming to the rescue; otherwise they would have kicked the fire away after giving their sacrifice a good warming. They probably had quite a lesson, having to worry for some time to come whether they would be brought to book for their mistimed practical joke. Anyway, our boys continued to play quite happily with the same friends afterwards.

Lin was expecting a baby in November, and her Mam was staying with us waiting the event, having been at the birth of each of our children. On Thursday, the thirteenth, Lin was in labour and I took her in to a private nursing home for the birth. Our daughter, Christine Mary, was a bouncing nine-and-a-half-pound baby, a big effort for my wife, who is of smallish build. When I got to the home after the delivery she had lain on the delivery table for a couple of hours, having refused to allow the nurse who delivered her child to transport her back to her bed using a "fireman's lift." I carried her back like a baby. An after-birth repair operation at a hospital later vindicated Lin's concern not to be woman handled in the manner proposed.

Shortly after the birth, Mam became anxious to return to the house she had left empty at Pembrey while playing her matriarchal part in Bridgend. We all decided that it would be better for Newton to stay with her for company after Christmas.

At the time I was travelling daily to Margam from Bridgend on an express coach, and one day I suffered the sort of indignity that can happen to anyone. I proferred the only money I had for my one-shilling-and-threepenny ticket, a one-pound note, with apologies for not having the correct change. Whereupon I was subjected to unseemly abuse from the ill-tempered conductor, who had already been berating other passengers.

"You're the one who is always giving pound notes," he raved. I should be so lucky! It was the first time I had asked for change on my bus journeys to and from work.

I protested and told him he was mistaken. If he had no change, he could hold the pound and take my name for me to claim the change from the offices of the company, but he would have none of it, stopping the bus for me to get off miles from my destination. Fortunately the day was saved for me when a good Samaritan from the rear of the bus came up and declared, "I will pay this gentleman's fare. I know him."

The conductor did not say a word, but grabbed the money, issued a ticket, and gave two rings of the bell to restart the bus. After that there was a distinct absence of conversation by the usually chatty passengers. When I alighted, my benefactor followed me. It seemed he was an employee of the Steel Company of Wales who had seen me often on site as I dashed around all over the place, getting lifts on site vehicles to speed my duties along.

When spring of 1948 came round, the situation of having my son in Pembrey while we were in Bridgend became unreal for Lin and me. We decided something had to be done about it. I elected to buy a car and travel from Pembrey again, relinquishing the firm's house in Bridgend to my colleague John Illingworth, who was the sectional engineer for the west section at the site at Margam.

The car I bought was a twelve-horsepower Flying Standard 1938 model that had seen better days in the hilly districts of South Wales. In no time it needed a change of engine, and for a time I was to experience the doubtful joys of secondhand motoring. Bill Beynon of A.C.E. garage Burry Port had his own business. He slaved away to make a go of the place, spending more time on his back under engines than any man I ever knew. It was to Bill Beynon that I went whenever things wanted doing to my car, and it was no fault of his that I spent a goodly chunk of my wages each month simply maintaining the car. The gear box arrangement on the car was a mock-up—a 1935 model box on a 1938 car—so a special bracket had been welded to the frame of the Standard to carry the longer box, causing troubles in the differential coupling in the transmission and in the gear box itself, which had been botched before I made the purchase. Bill persevered many times to effect a repair before a solution came that I shall describe in the next chapter.

The day came when I saw an advert in the *Western Mail* for a temporary resident engineer vacancy in the Water Department of the county borough of Swansea, offering nine months' work at a salary slightly larger than I was getting. Swansea being a few miles nearer Pembrey, this position was of interest to me, and I mentioned it to my friend John McCallum. He knew what the job was, having worked on it for the contractors, none other than Geo. Wimpey, and was able to fill me in with a few tips of the salient points relating to the service reservoir being built at Cwmdonkin in the borough itself. Fortunate to be short-listed, I turned up at the guildhall in Swansea resplendent in a new, clerical grey flannel two-piece suit especially bought for the interview. I was third out of three to go in, and from the confidence of my rivals I felt

I was just making up the numbers. When called for the interview, I had already gleaned from the applicants who had preceeded me roughly what was the layout of the interview panel.

When I knocked at the door and entered when bade to do so, the interview had a sort of inevitability about it. As I strode into the room towards a top table, with two wings like it had all been set out for an after-work beano by the Committee of Councillors now in session, I returned the chairman's "Good afternoon," with warmth and interest that so many people should be prepared to attend to ensure they got the man they wanted. It was as though the candidates were to be examined like a biological specimen in a bottle, from all sides, as the chair for my use was placed in the middle of the space enclosed by the tables.

The chairman was elderly Alderman Henry, and he was flanked by Trevelyan Price, engineer and manager, on his left and by Leslie Evans, the deputy, on his right with sundry others completing the seats at the top and side tables. Perhaps I had a swivel chair, perhaps it was fixed, but my memory is of my responses being made as though from a man pleading his case to avoid being sentenced to some dire tribulation for failure. It was the first time I had been interviewed by the heavy mob of a local authority. I was thankful for my preparation and tipoff on the nature of the job by McCallum, as the analogy of the work they had in hand with the soaking pits structure at Margam was invaluable in helping me to keep the interview flowing, holding the interest of the gentlemen trying to get to know me in the limited time they had set themselves. When I left the room, I felt I had enjoyed the experience, without really caring whether I had been successful or not.

When Les Evans came into the anteroom and asked me to return, I assumed in my ignorance of how these things were done, that I was being given the bad news of being unsuccessful. When I sat in the chair they told me I was being offered the post and begged me to make as early a start as possible. The nine months' job to finish off a service reservoir was mine.

Success in landing a job that one has looked forward to is a sweet feeling. Back in Margam the next day, I went straight to Dick Wright, the agent, before mentioning to anyone else my departure, in four weeks' time to start at Swansea as a resident engineer for the Swansea Corporation Water Department. Dick was still easy to talk to. The loss of yours truly was not the end of the world, as floor finishings were in hand in some of the sheds and the main heavy foundations were well on for most of the work currently in hand at the site by Wimpey, the main contractors.

My imminent departure from the mammoth works and my colleague was now tinged with a little sadness.

With my departure early in August 1948 came the beginning of a new episode in my life that would last not for nine months, as I anticipated, but for nearly eight years

12.   Drops of Water

Maj. Norman S. Williams had left his job as the resident engineer supervising the Cwmdonkin service reservoir construction a month before I arrived. For his part he had landed a plum job as R.E. in charge of the construction of the new Uskmouth power station, an £ 18 million project. In the interim between his departure and my arrival as his replacement, the Swansea water engineer and manager, Trevelyan Price, sent out John Gamblin, a young engineer from the department's office in the guildhall, to keep a check on lines and levels and generally hold the fort. John seemed glad to see someone arrive so he could return to his work in the guildhall.

I was glad to have made the change even if the work prospects were only for nine months' duration. It meant that I would now be a big fish in a small pool instead of a small fish in a big pool, and with the new job concerned being a service reservoir, just a covered reinforced concrete water tank of the then value of forty thousand pounds, how apt that comparison is. I was just hoping that I might now have a future, with the pools getting bigger for me to grow with them, although I knew of no future with Swansea at the outset of my time at Cwmdonkin.

Cwmdonkin Terrace gave access to the entrance of the works. It was a street of houses on the top side only, looking out from its elevated situation across the beautiful vista of Swansea Bay. Below the terraced road the land sloped at about one in five and the 1.5 million gallon storage twin tank reinforced concrete structure had its foundations half sitting on hard underground rock where the excavation went into the hillside and half on mass concrete with big stone displacers, the most sensible way of making up the level, as the top of the rock went lower down the slope of the hill.

The reservoir was part of the postwar improvements for the department that had already been carried out by contract on the Morriston side of the second most populous municipality in Wales, with the new Morriston pumping station and pipelines and the new Penlan water tower parts of the Clase Water Supply scheme.

On the day of my arrival I was plunged straight into the role of 100 percent of the council's engineering staff. Bill Dalton stood there, out on the works below the office, in his capacity of 100 percent of the clerks of works, and in my office, the portable wooden-hut type, sat Fred

Greenway in his role of 100 percent of the administration staff, i.e., the R.E.'s clerk. To complete our team, Andrew Donald kept bobbing around between Bill Dalton on site and Dalton's own office, a few feet from mine, where he made an appearance several times a day. Andrew's job was offices' attendant and chainman. He proved to be something of a comic entertainer, especially when attempting to light the Tortoise type coke-burning stoves. He would dash from office to office, dropping in the coke first, sticks on top, and then the paper, soaked in paraffin, with a lighted match thrown in, nearly causing an explosion but seldom getting the fire to light in the upside-down technique he had. Fires were Andrew's bête noire.

When I first asked him his name, he answered, "Donald, sir, Andrew Donald."

"And what do we call you?" I pressed.

"Either name, sir," said Andrew. "They're both Christian names, Donald and Andrew."

Edward Davies was the contractor's engineer agent, and for most of the time he was 100 percent of the engineering personnel. Harry Newport came from Newport, and he was 100 percent of the foremen for the contractor, Geo. Wimpey, whose area offices for such works were in Cardiff.

When I first arrived, it was to the happy sight of the July monthly certificate application for my checking and onward transmission to the guildhall. The work was at a stage where one could look around and measure up without the embarrassment of having to countenance work covered up before I had seen it going in, so apart from a quick perusal of the documents and drawings and an examination of works progress with my clerk of works on site, I was not under much pressure, so that the contractor's interim application for money was not held up long by me.

Personally checking lines and levels was a refreshing revision for one who had acted in executive jobs since promotion from Fishguard to Liverpool seven years previously.

I did not see the water engineer and manager for a month, though I visited his deputy, Les Evans, several times at the guildhall, in this period preparing a statement of the additional items to warn the corporation of some inevitable costs not provided in the tender sum. As a result, the contractor did not need to beg cap in hand for due entitlement and, inter alia, he did not find himself in the position of having to claim for more than he was entitled in order to settle for what he was entitled, as is often the case in these areas of negotiated settlements of final accounts of civil engineering works. I found Davies was a genuine man to deal with as, in fact, most men are when trust has been established.

After a month, Trevelyan Price (no relation, as the borough had announced when they published my appointment in the paper) visited the site, and he looked in regularly after that. He was a rotund, jovial-faced engineer who sped round his territory in a Rover twelve-horse-power car, DCY 700, a Swansea registration number. He had had many years with Swansea, working his way up from a junior. Old Price was pleased with the way the reservoir was progressing, and apart from two minor contretemps I would have with him in the future, he and I usually hit it off very well.

The routine of my new job kept me with my nose to the grindstone, which was very refreshing. Like anyone involved in the practical details of smaller works, I was neither bored nor had I time for illusions of grandeur.

On the family side, however, one sad thing happened to us. One Sunday morning Lin's mother, Old Mam, was talking to me in the garage I had recently erected near Cliff Cottage with the help of friendly neighbours from the Furnace, Pembrey. She gazed admiringly at the sound corrugated aluminium structure with its clean-cut duralumin frame sections and well-hung doors, a garage I had bought from Western Disposals, Newport. Then she looked at the pink floor I had laid for the garage, cementone coloured concrete, the aggregate being bricks broken down small with a lump hammer over a period of weeks.

"It's like a room, Jim bach," said the old dear, her cheeks glowing with pleasure.

We went into the house. Lin and I then went out for a few minutes of fresh air and came back after no more than ten minutes after a walk to "the burrows" on the links, the Ashburnham championship golf course that flanked the Cliff garden with its fairway for the first green. Mam was sitting in her chair in the sitting room, breathing heavily. We could not waken her. I got in touch with Dr. Leslie Williams, our family doctor. He came straightaway, but there was nothing he could do. "It's the hardening of the arteries," he explained.

This was the end of an era. Lin had lost her father in 1935, and now, thirteen years later, her mother passed away. It was good that I had not stayed with the Admiralty, good that I had finished with Margam, good that I was working at Swansea, but these thoughts were just palliatives to ease the pain of that day. My father and mother came down for the funeral. They had only been down once before, when our third child, Linda, had died in October 1944 at the age of sixteen months, as I recounted in chapter 7.

My Standard 12 had seen its best days and was costing me a lot in repairs, apart from the time wastage in having to journey into work by walking to Burryport Station, training to High Street station, Swansea, and then bussing up to the Uplands, an overall journey of about nineteen miles each way.

The main problem was still the gear box, and I must have had it repaired at least three times at a cost of over two weeks' salary each time, when I was blessed with good fortune. There was a car like mine belonging to the department, a Flying Standard 1937 model, just sold by contract to a man who did not bother to collect it, crying off after making the deal. The car was a writeoff, only sold for scrap, but it had some reasonable parts in its eleven-year-old shell. Thinking of my gear box, exactly the same type as that of the scrap car, I put in the same bid, and finally it was accepted, so there I was having it towed back to Pembrey by a breakdown merchant, steering my new acquisition with feelings of yesteryear when I had been involved in towing for my father, the caravan chassis and the Graham Dodge lorry, which I described in chapter 3. When I arrived at the house in came another Standard twelve-horse-power car behind me, the engine switching off at the cottage gate. "I saw you going through Pwll," said the driver. "Would you care to sell me the carburettor?"

I agreed and asked him for fifteen bob. The money was hardly in my hand when he had whipped off the carburettor and was on his way.

That was the start of a chain of petty sales of spares off my old car that redressed the losses I had sustained in repairs with a little to spare. Spares were so hard to find at a time when no one had a new car and the old, black wreck of a car (all cars were black in those days) was soon down to its carcass. I had already sold the engine block three times to friend Beynon to carry out reconditioning work of replacement engine purposes. Then along came a mechanical engineer friend from the R.O. factory at Pembrey, with his own Standard twelve-horsepower body eaten away by the salty air and the fumes from the sulphuric acid plant in building no. P3. He jacked up his rotting car body from the limb of one of the stout trees in front of the Cliff and exchanged it for the body of the former wreck, which, compared with his, was quite respectable. Finally, along came a rag-and-bone merchant to remove the rusted carcass left behind by my mechanical friend.

It had given me quite a lift to recoup my losses on the car maintenance and repairs, and my travel to Swansea in the car did not give me undue trouble for the next two and a half years. Back on the site of the works, construction was going along nicely. By the late spring of 1949, the walls and columns were finished and scaffolding to support the roof construction was being erected. To supply this service structure, which was intended to offer three days' storage reserve for the area it would supply, pipe connections had to be made into the complex mains system at the nearest street intersection. This would be fiddling work for a huge firm to tackle, as such a firm tends to be departmentalised and needs plenty of bulk to get its teeth into in order to justify its overhead costs. Wimpey elected to pull out of the job after getting the bulk of the work finished, so they came to an arrangement with the borough to complete all but the pipe connections in the town street lower down the hill, this work to be done by the Water Department personnel, more at home with this type of work. However, while the firm was still engaged in laying the straight runs of pipe on Cwmdonkin Hill, one pipe, weighing about a ton, started to roll down the hill.

The general foreman, Harry Newport, was a big man, and he tried to catch hold of the pipe and spin it into the verge to bring it to rest. Harry was big but the momentum of the pipe spinning round like the wheel of a car caused Harry, who hung onto the end of the pipe, to spin up into the air and perform an unintended somersault worthy of a champion gymnast. He returned to the office a sadder, wiser, sorer man. His instinctive reaction was the same that could be expected from many a man, yet the consequence on the man was the same that one would expect of someone catching hold of the tyre of a car running down the hill, and no one would dream of doing such a foolhardy thing as that.

Eventually with summer upon us, July saw the roof of the reservoir concreted, the water testing in hand, and the soiling over the structure done, with seeding of the soil and all other surface works completed by August. Davies and I had no difficulty agreeing on measurements for payment of the work, and I was ready for my next job, which—surprise, surprise—came with the offer of being resident engineer for a twelve months' scheme for the Water Department of the county borough of Swansea.

My second job with Swansea was for the supervision of a contract let to Hussey, Egan, and Pickmere for the provision of a new service reservoir on Clyne Common to act as a reserve for the developing West Cross area overlooking the bay. The capacity of the reservoir was three-quarters of a million gallons, only half that of Cwmdonkin, and it was to be a circular structure, not a rectangular one. Pipelines were to run from Black Pill, on the Mumbles Road, near the shoreline of the bay, some two and a half miles away from the common, the new lines having to pass up the Mayals and cross the Clyne Golf Links, which was inte-grated with the common.

I was pleased to consolidate my experience in this class of work, which kept me fully occupied. I was also grateful for the honorary mem-bership of the golf club arranged by Trevelyan Price, but unfortunately, I never got around to playing the game, although I always carried in the car boot the bag of assorted secondhand clubs I had purchased in Singapore for eleven dollars (£1 : 3s; 8d) apiece.

There was the problem of getting to work each day at the site on Clyne common, remote from my home in Pembrey, a tortuous twenty-five-mile distance. With no car allowance, I was torn between the alternatives of the expense of motoring and the relatively cheaper, more time-consuming complicated journey of walking the mile to the station, taking the train, then the bus, finally legging the last half-mile across the common to the office. Usually I took the car, as I watched the money I had saved in the Far East dwindle slowly away.

The whole scheme had to be completed in twelve months, and that is not a lot of time for a scheme of its kind, allowing for factors like the vagaries of the weather and, in those immediate postwar days, shortages of materials and experienced labour. It was essential to tackle the construction on two fronts. Don Brown, Hussey, Egan, and Pickmere's area agent had the round hole excavated for what would become, when restored, an unobtrusive, flat mound at the highest point of the common and over the big vessel storing the elixir of life. At the same time, four classes of pipes were being offloaded along the line of the water mains, stacking on the verges in the Mayals or strung out on the course heathland of the common, where a rough track existed for a tractor to travel up to the golf club maintenance compound, nestling discreetly below in a hollow, below the general lie of the land. The hollow was probably the result of digging soil to make the platforms for teeing off at each hole on the course. When I refer to different classes of pipes, I do not mean they had different diameters, but different requirements according to the pressure of the water at a particular point. Thus down at Black Pill far below the surface of the water in the new reservoir, the pressure of the water would be greatest and steel pipes capable of withstanding a test load of a thousand feet head were installed, while up at the reservoir end, where the pressure was lightest, asbestos pipes were adequate. In between, spun-iron pipes of intermediate strength were employed, with special adapters for coupling one kind of pipe to another. All iron pipes have some kind of protection against the ground and against the water they will contain, but asbestos cement pipes did not require such protection in our situations; therefore, the choice of pipes is one that requires careful attention.

The numerous small quantities of work to construct a reservoir of this kind meant that the contractor's supervision of the gangs and our own supervision were continual. Thankfully, my staff was still the same as I had at Cwmdonkin; Bill, the clerk of works, Fred, the R.E.'s clerk, and, in lighter, still essential, vein, we had dual nomination Andrew Donald still there to act as general factotum in the field.

As the work progressed, the excavation of the chamber was tipped into the old quarry holes nearby, a thin layer of concrete was placed as oversite blinding to keep the floor of the excavation clean prior to placing the reinforcement, and timber work and concrete work for the floor of the structure were all constructed in bays, as is the usual practice, checkerboard patterns and water bars also being usual to minimise shrinkage as the concrete hardens and to obviate leakage. The same precautions were followed in erecting the walls. Everyone concerned with the construction was aware of the stringent test that would have to be withstood when the tank had been finished, watertightness being proved for a whole month before work of backfilling with soil would be allowed around and over the structure.

Inside, the smooth, dense finish of the walls is an essential requirement to ensure bacteriological growth is discouraged. No quarter was given and none sought of any porous spots visible on the walls. These had to be made good properly.

While the reservoir was proceeding, the pipelines were being laid and tested in lengths of about one hundred yards at a time. Such tests were prescribed according to standards laid down by the British Standards Institution, and I always made it my business to be present at the tests. At the sharp end of testing you should always see an engineer. His training is vital to assess the integrity of what has been constructed.

One day there was a little light relief when in came Dick Lillicrap from the guildhall. He had recently been made the deputy for the engineer and manager, having been upgraded to this post when Les Evans was appointed water engineer and manager at Epsom and Ewell in Surrey. Dick was a lively one and was trying out a newfangled vehicle designed to run on rough country. It had the Swansea registration number GCY 21. It was square and green with a canvas hood and carried the spare wheel on the flat bonnet. It was, of course, one of the earliest Landrovers.

Dick breezed into my office. "Care for a ride?" he asked.

"Try anything once," I replied, joining him in the front seat.

"Hang onto the bar," he advised.

Then off we went to test the vehicle with its four-wheel drive and its transfer box to give it eight forward gears and the choice of driving on two wheels only. Dick seemed to think he could balance us on two side wheels at a time, and I was both thrilled and concerned for my life as the fiery Dick, three years younger than I, put the vehicle through its paces. It was a trial run for the model, which the department were purchasing for use on the Usk Reservoir scheme up in the mountains near Trecastle.

No cowboy in a rodeo had anything to show Dick that day as we dived down out of sight into the old quarry workings and came up again like dolphins in the Mediterranean Sea, while Dick held the wheel and I feared the bar on the dash would come loose in my hand. The Land-rover was some vehicle, some toy for a man to take delight in, and I could well imagine its possibilities as I looked forlornly down the rough track up the common where I had been obliged to leave my car half a mile away, lest the innards fell out in negotiating the rocky way to the office.

When the Clyne Common job was completed, only five claims were raised by HE and P, and in a meeting with the water engineer and manager and Gerard Egan, my report, regarding proportions that fairly represented elements provided during construction but not allowed for by the original contract, was accepted without demur on behalf of the parties to the contract, and I had the satisfaction of knowing that I had made a substantial contribution to providing a few extra drops of water for the people of Swansea.

Added to this, I was now required to take on a new contract, for the construction of the Clase water tower, for which a contractor had already been selected by competitive tender. Many's the time I had looked at the Penlan Tower, high on the landscape, as I had driven into Swansea and pictured myself the resident engineer for such a structure, so the news was just like having a dream come true.

Bill Dalton, my clerk of works, a permanent servant of the department, had only been coopted into the contract work on a short-term basis. He now clicked for the job of superintendent. This was good for Bill, sad for me, but I had the satisfaction that Fred, my clerk, would remain with me on the venture. Fred's presence always brought some light relief to my daily routine. He was a man of small stature and big heart, uncommonly thin on top, with a thin wisp of hair brushed over to retain the last vestiges of fading youth, he being a mature family man in his midforties.

Two repetitive quirks have implanted themselves in my mind. When Fred had to attend to the calls of nature, it was necessary for him to cross over from the office, where he toiled away at typing or checking figures, to the chemical latrine privy a few feet away. Fred would always put on his hat to make this journey. Perhaps he did not wish the three strands of hair to be disturbed. Perhaps he was afraid to catch cold. Perhaps he was self-conscious. I may never know, as I was too polite to make a pointed remark to query the reason of my good office colleague.

The other firm recollection of Fred is his oft-repeated account of his regular visits to the Vetch Field to watch Swansea Town playing soccer. It was not the account of the soccer itself that preoccupied Fred, though the name of Ivor Allchurch did crop up quite frequently. It was the description of "our Arch." Arch was Fred's brother. The diminutive Fred spoke proudly of his brother, who from all Fred said must have had the physique of "Big Daddy," at least in Fred's mind. "Guvnor," Fred would tell me ad nauseam, "I'm always glad that Arch comes with me to the Vetch. When I'm in danger of being crushed, Arch just *leans* on the crowd and gives me a clear view of the match in safety."

The two idiosyncrasies were to be solace for me for another year, and I was looking forward to what was in store for the next twelve months.

A new face took the place of Bill Dalton as clerk of works. It was Sam Williams, at sixty-three an old-timer, who had spent his life in Swansea and worked several years in the water industry, with recent years on pipe lines in the Gower, and most recently he had worked for Swansea on the supervision of the construction of the Penlan water tower. Sam was an old-fashioned fag-smoking, earthy citizen of Swansea, with his home a cock stride away at Treboeth. He was heavy-jowled, wearing a flat cap, baggy jacket, and trousers.

The water tower was to be constructed of reinforced concrete and the 150,000 gallon circular tank, seventy feet up in the air, would project out with a splayed bottom oversailing the supporting structure below, a cylindrical wall stiffened with outside piers. From the drawing it appeared like a neat button mushroom. The site was on high ground at Treboeth, a district on the Morriston side of the town and well chosen from the point of view of a solid foundation. At the prescribed depth, decided after examination of the ground condition in trial holes, firm rock with horizontal bedding made perfect support for the heavy structure, and no unforeseen problems were encountered in that connection.

With the solid reinforced concrete raft-type foundation in place, the stout walls and piers went up using rigid arc-shaped sections of strong wooden forms lined with aluminium sheet to give the concrete a smooth finish and to enable the "shuttering" to be used several times. Once the walls had been built from the foundation up to ground level, backfilling around the wall was well rammed into place to ensure the backfilling would withstand, without settlement, the weight of the outside scaffolding necessary to erect the high structure above ground level.

There were vertical slit window openings left in the walls as they proceeded upwards, and at the top of the wall a strong ring beam of reinforced of concrete had to be poured.

When we were about to construct this part of the work I was struck down with the flu and confined to bed. However, this did not mean that I would be left in peace to forget the work and sweat out the feverishness. Old Sam was a cautious man and was conscious of his own limited responsibility vis-à-vis the contract. He had a niggle about the manner in which the contractor intended to construct the ring beam as laid down by the contract particulars, and so he turned up at my house, plan in hand. "I think we should do this 'in one,' pal," he said. "It would be fatal to do it in sections."

I saw the point through my aching, ill-focussing ocular organs. It was my prerogative to decide such a matter "by virtue of your appointment," as Trevelyan would have been only too glad to point out. The change in method would be marginally beneficial in that it would minimise room for poor workmanship. It would also have the confidence of my clerk of works, who was expressing his gut feeling in the matter, so I agreed to the change being made. Sam was delighted to return, knowing that his journey had been worthwhile, and with a Parthian shot as he left my bedroom he declared, "Make b—— sure you stay in bed till you are well over that old flu, boy." Sam was just one of many such stalwarts who grace the field of public works activities in our noble land. We hear of Brunel, Telford, and McAdam, but without the Sam Williamses of this world they would have remained anonymous. They are the N.C.O.'s of the construction armies in our civilian life.

The London firm A. Jackaman and Son Ltd. was building the water tower, and the son, Nigel Jackaman, made frequent visits to the Swansea job to see how things were progressing, always observing the courtesy of calling in on me to have my comments or complaints. It was patently obvious that the firm's outlook was to give satisfaction before exacting payment for work.

Nigel Jackaman was a good man to deal with, not given to "fly" arguments or leading questions. The genuine approach to work was also apparent in the quantity surveyor sent to site to draw up the certificate applications for payment every month. Nigel was as British as his Rover saloon car.

I ought to, perhaps mention a more social occasion that had taken place on February 16, 1950, while I was busy with the supervision of the Clyne scheme, then about halfway complete. It was an occasion to mark the opening of the Clase water scheme or, rather, the main parts of it and also take in a tour of inspection of the Cwmdonkin Reservoir and the works in progress at Clyne scheme. The Clase scheme was for improvements of the supplies distribution system to meet the postwar housing construction, with projected development of large areas of land at Clase, Penlan, Gendros, and Forest Hall, and entailed new pipelines with service reservoirs, water towers, and pumping stations strategically sited in proximity to the new developments.

Present at the special occasion were the then mayor, Councillor R. Gronow, J.P., his deputy, Alderman W. George, J.P., Councillor R. Henry, the chairman of the Water Committee, the other members of the committee, and officials of the borough, including the town clerk, T. B. Bowen, the water engineer and manager, Trevelyan Price, and other personnel of the authority. The contractors concerned were represented by people like Gerard Egan and Nigel Jackaman, principals in the respective firms of Hussey, Egan and Pickmere Ltd. and A. Jackaman and Son, Ltd., as well as senior employees responsible for their work in the field. Normal Williams and I were accorded generous thanks by T. P.

It was on an impulse when leaving Clase Tower site on the day before the Derby, 1951, that I warned Fred not to expect me in the next day. I suddenly decided to go to the Derby. Epsom and I were strangers. The only connection I had was Les Evans, and I was unlikely to bump into him in all that crowd, even if he were numbered among the madding throng. On the other hand, it was impossible for anyone to attend the races without bumping into the famous racing tipster Prince Monolulu, resplendent in all his regalia, colourful, barefooted, befeathered, and authoritative. "I gotta horse," he kept bellowing to all around. People were buying his small envelope-type slips as if they forecast the date of the end of the world.

Royalty was represented with Queen Elizabeth, the wife of George VI (now Queen Mother) and younger members of the family. It was the same sweet face and charming manner that we have all loved for donkeys' years now that TV has taken over our lives. "Do you?" I caught the words she uttered, enough to convert me to her fan for life, and she glided off past me, standing among those in the railed off crowd, her entourage including the young princesses. There was a tremendous atmosphere on the public field, which was cut up under the heavy use by vehicular and human traffic.

My bet on the Derby was not successful. I cannot remember which also-ran it was among the field won by C. Spares on Arctic Prince. I was out of pocket but in on experience, and as I contemplated my experience at Epsom, my work at Clase was sweeter from that day until I saw the job finished in the August.

Back in the guildhall, my presence was required to comment on the validity of a few claims, and after some deliberation, the meeting ended. I bought a paper and saw that a vacancy had arisen for an R.E. to work on the Usk Reservoir scheme, the advertisement having been placed on behalf of the consulting engineers for the scheme, Binnie, Deacon, and Gourley. With no promise of further work in the department, I slung in an application to the consultants for the big scheme in hand for the corporation.

Final dotting of the i's and crossing of the t's regarding the settling up of the tower contract was almost done when I appeared for an interview at the reservoir construction site at Trecastle to meet the engineer responsible for the scheme at the firm's London office.

George Arthur Reece Sheppard was a full partner in the consulting firm and was at the site to see me for himself, as an essential part of the screening process before taking on a prospective appointee. It was shortly after our meeting that I was appointed to the post of resident engineer (pipeline and treatment works) for the scheme, which would double the supply of water to Swansea.

The total length of trunk water mains from Myddfai village to the terminal reservoirs in Swansea was thirty-five miles, with a water treatment plant to be built at Bryn Gwyn, ten miles down the line, near Llandeilo, and a break pressure tank required on the Graig Fawr mountain near Ammanford. Associated supply contracts for the pipes and valves would also come under my purview as the resident engineer for this very substantial part of the overall Usk Reservoir scheme.

It was a move to a bigger pool, fulfilling the promise of my aspirations when I first came to do the Cwmdonkin job, though I had not thought I would be fortunate in gravitating to such an opportunity with the same employer. If I kept my head down and did not get caught on a hook, I might well regard myself as becoming a somewhat bigger fish in the bigger pool.

"Onward and upward," said Trevelyan in congratulating me on getting the job. I would still be paid by the borough, but would be subject to the technical guidance of the firm that had designed the works and retained overall responsibility for the works until completed, commissioned, and maintained for an appropriate period, which was a further year after construction.

For the next four years and four months I was to have the most interesting, most exercising, most physically and mentally stimulating challenge of my life. This was the first really solid major public works scheme for which I had, to all intents and purposes, fully delegated site responsibility. Although I believed I had the ability and many precedences of experiences of allied works, I knew I had to put it all together to see this big new job through to the end.

When laying trunk mains through hostile terrain, not that we would encounter any red Indians, we would have to see that every one of six and a half thousand twenty-six-foot-long steel pipes would be properly located and laid in outlandish, sidelong hilly countryside, bedded and covered over, each pipe length a separate job in itself, yet just a minute part of the total work to be done.

What was life like in those four years? One of the first things I did after being appointed to my new job was change my car, now thirteen years old, for another one, a sporty 9 HP Singer Le Mans. That was a

Swansea Waterworks (1950–55).

mistake if anything ever was. My enthusiasm waned when, as the weather got cooler, the fresh air coming up through the floor boards around the clutch and brake pedals was far too excessive for comfort. Also, the two large doors, which opened outwards from rear mounted hinges in the light aluminium body, constituted the real danger of depositing myself or my passenger on the road. Experiences of this possibility with L. B. Aylen, the chief resident engineer, at the established site offices for the reservoir and also Lin caused me to become disenchanted with the car within a few weeks of purchase, so I replaced it with a 1938 model Rover fourteen-horsepower saloon. This was more comfortable and acceptable to my wife, but took a substantial share of my salary, as no car allowances were payable, site transport being available in the form of Landrovers.

In August 1951, pipes started to be delivered to the site and the contractor, Richard Costain, installed an engineer, B. Kearsey, to attend to the receipt of the pipes and undertake some early work on preparation, with the construction of some of the piers for pipe crossings, prior to the arrival of an experienced Agent. During the first weeks when this sort of preliminary activity was in hand, Kearsey and I, having no offices on the site, were obliged to work from our cars, leaving notes for each other on the underside of a bush in a rendezvous rather in the fashion of Soviet spies. We were then able to go our separate ways to perform our respective duties, doing the work for the contractor or inspecting and certifying it on behalf of the corporation's consultants.

Mobile cranes were offloading steel pipes of 28 3/4 inch outside diameter, each twenty-six feet, two inches in approximate length, having to be measured to within a sixteenth of an inch of accuracy for purposes of checking the bills submitted by the suppliers. The pipes were being stacked in locations conveniently close to the line of the pipeline, over the laneside hedges along which the cranes could travel, pipe lorries collecting their loads from the railway trucks in the Railway Goods Yard in Llandovery.

We sustained many scratches while squeezing ourselves between the hedges and pipes, which we had to inspect at their ends to record each reference number, an essential but tedious chore. It is true to say, however, that what seemed to be a bit of a bind became a source of great satisfaction when the accuracy of the accounts being presented by the suppliers was thereby established. There was another problem associated with the deliveries, and that was the amount of the counter charges in respect of repairs to the bitumen lining and sheathing, where damaged in transit on the railway, the point of takeover being when the goods were offloaded from the railway trucks. Inspection therefore had to be made instantaneously when the pipes were suspended and capable of examination. For several months Kearsey and I persevered to ensure justice was done vis-à-vis the pipes themselves while also setting out the route for the line of the pipeline across the fields, the hillsides, the marshes, and woodland, valleys and streams. The line was established all the way from its source emerging from the mountainside way up the valley above Myddfai village, down to the site of the treatment works at Bryn Gwyn, and up the hillside east of Llandeilo, towards the Black Mountain.

Many of the concrete piers for crossing valleys were being constructed at this time until presently, after the initial months, the site offices were being erected at Bryn Gwyn. Pipelaying gangs and excavating machines began to arrive to tackle the main object of the exercise, the pipeline itself. The easement width of twelve yards had to be stripped of its topsoil or cleared of its woodland or other growth, all soil being placed at the edge of the easement in the form of a windrow to be carefully preserved for returning to restore the disturbed land.

As soon as the contractor's nominated agent arrived, work began in earnest laying pipes. Men and staff were soon increasing in numbers to accord with the progress of the work.

A memorable incident occurred as the contractor was trying to clear trees from the wooded site of land on either side of Nant Alus. About the beginning of November 1951, Kearsey and I were apprehensive over obstruction from a farmer on the north side of the Nant whom we had nicknamed Persil White. This was because as we were pegging out the line on the south side approaching the valley for purposes of site clearance, the farmer stood on the boundary of his farm on the north side with a dog at his heel and a double-barrelled gun under his arm, looking across the valley, conspicuous in his wide expanse of white shirt.

We were very conscious of the pointed interest the farmer was taking in our coming and too wry to venture onto his land when he had us at a disadvantage. Kearsey came up with an idea for procedure on the north side. We agreed to approach the north side of the valley from the north of the wood, which lay beyond the field on which the farmer was wont to stand on sentry. I implanted range rods to line in with skyrockets let off by Kearsey on our last peg, south of the valley. We thus had the line for the big D7 bulldozers to press on like Sherman tanks from the north side in the direction of the valley, uprooting the trees of the woodland to claim the easement width.

The opposition we feared was probably more imaginary than real. At any rate, it was not forthcoming. It would have had to be very deliberate to stop a D7 bulldozer. One can pinpoint the time of this event as early November 1951, as the skyrockets were in the shops for Bonfire Night.

To be perfectly fair, considering that the pipeline scheme, drafted some twenty years previously in 1931 did not have statutory force so far as alignment was concerned, such variations in line as were expedient to avoid bad ground or other reasons were tolerated with the patience and cooperation of the farmers whose land was cut through by the new trunk water main. Take old Jones of Cruglas Farm, for instance. He used to come up from this house a few hundred yards down the sloping fields below the line, to have a good look at the pipes in the trench, central in the wide swathe of land disturbing his land. Now in his seventies, he had had wind of this phenomenal happening for two decades, a very long time for someone to live with such an event looming up.

When the work began to open up with the arrival of Costain's agent, our respective establishments based on the site offices at Bryn Gwyn grew accordingly. Landrovers were provided by the contractors for their people, as they were also in the case of the Swansea people. Opposite numbers were emerging to support the agent and myself, Joyce having his senior engineer, Kearsey, with juniors and his general foremen, and gangers, while I had the help of my assistant resident engineer, Keith Lewis, my chief clerk of works, Ernie Jones, junior engineers, generally undergraduates or graduates from Oxbridge universities under article agreements with the consultants, and several inspectors posted along the line to watch over the work wherever the individual gangs were operating.

I shall ever feel indebted to Ernie, a man of exceptional qualities as an inspector, with unbounded loyalty, great energy, and a compulsive drive to ensure that everything that was done on the job was done properly. He lived with his wife and pet dog in a caravan in Llandeilo. It was a daily ritual for Ernie to accompany me on my round of the work sites, seeing the pipelaying actually being done, length by length, the testing of completed lengths of a mile to a mile and a half between control valves, and the erection to the piers in the rivers, like the Sawdde and the Amman, calling for careful attention of construction of foundations in fast-flowing waters. Minor streams flowing down the hillsides had to be taken under or over the pipeline, as the nature of the ground dictated where these intersections were encountered. It was usual for me to find that my inspection round had involved me in seventy Landrover miles, a good proportion of which involved travel over acutely sidelong ground, having a side slope of up to one in three and a half, along the contour of hills and the mountain of Graig Fawr in the area of the Black Mountain.

We would return to the office at Bryn Gwyn, and Ernie would go off again to keep the pot boiling in supervising the work, overseeing his inspectors at each point, liaising with the graduate engineers gaining experience while they did their engineering duties on the sections allotted to them. Meanwhile I would be in the office attending to contract correspondence, dictating instructions, talking on the telephone, or co-ordinating the work in the drawing office and maintaining a discipline to ensure a smooth flow of work to maintain the progress was compatible with the expectations of programme, specifications, and the budgeted costs. Fred Greenway, still with me to take charge of the office management, now had the assistance of Grace Williams, a school-leaver, daughter of a Llandeilo bus services owner, Llewellyn Christmas Williams, whose natural bent as a commercially minded person made her of use more than average for her years.

In due course a contract was let for the treatment works at Bryn Gwyn, a rapid gravity filtration works to filter and purify the 8 million gallons of water per day designed to discharge down the pipeline and double the available consumption of water for Swansea, where major works like Velindre and Trostre Steelworks were projected. My new assistant resident engineer, Clifford Haig Turner, was of invaluable help in supervising this part of the works. Cliff Turner showed an intrinsic interest in all that he had to do and was indispensable to me in dealing with the agent for the works, Mr. Parry, also of Richard Costains, who had been successful once again in getting this contract by competitive tender, ensuring they would be the main contractors for the whole of the Usk Reservoir scheme.

This new construction contract also included the Break Pressure Tank to be constructed on the steep side of Graig Fawr mountain, two miles away from the centre of Ammanford town, which lies between Llandeilo and Pontardulais on the way to Swansea.

Months had become years as the pipeline moved inexorably up hill and down dale, the gangs overcoming the obstacle of the hostile terrain with ingenuity and brute force. Joyce was replaced by Donald, who was able to infuse a greater sense of urgency into site activities. During his tenure as agent there was a very creditable buildup of progress, the aggregate of pipes laid peaking at sixty per day. This was a great period when I was beset with a challenge of keeping the laying contractor supplied with pipes by the suppliers at a time when steel authorisation certificates introduced during the war were still adopted by the government for discipline in the engineering and building industries.

There was the day when Donald sent me a letter to say that if pipes did not arrive on one section, there would be a claim for being held up. He intimated that the work there should be instructed to stop so that the gang could be moved to another area where he had pipes strung out. It was a ploy for him to avoid tackling the soft, marshy area ahead, where he admittedly had few pipes strung out but where, it seemed to me, he was not likely to move at a rate of knots. Tongue in cheek, I replied, "Carry on with the section, on the assumption that you have the pipes on the site in that location." Well, there was no answer to that, so I felt my duty had been done in avoiding unnecessary cost on the part of my masters. In this class of work, as you might imagine agent and resident engineer both have a duty to their respective firms and, so long as the game is played fairly without undue pressure from above on either side, there is much to be said for the policy of every possible cooperation.

As the pipeline proceeded relentlessly down to Swansea, where it had to connect with new terminal service reservoirs on the outskirts of the borough, it became necessary to establish a sub office in Ammanford town. By now, Fred Hoare was contractors' agent for the pipeline and the lower section of the work was carrying on round Graig Fawr and passing Pontardulais in the direction of Llangyfelach, where it would bifurcate to run on in two directions in order to supply the new service reservoirs at Clase and Cockett.

The year 1955 came and with it a dry spell to make it imperative that the reservoir at Trecastle and the supply mains, including the essential treatment works and break pressure tank, were completed and commissioned before stocks of water were exhausted in the existing reservoirs supplying Swansea water. The new reservoir was to be opened by the queen and Prince Philip on August 6, 1955, and my staff were anxious to be invited to the official opening.

David, Eluned, Christine, Jim, and Newton on the beach at Sully, Glamorgan, during their summer holiday in 1954.

There was some coolness over their invitations, so I had to go personally to see the town clerk, T. B. Bowen, as so often with these occasions the people who are at the sharp end of actually doing the work are thinly represented. When we did get our invites and sat there in a half-empty covered stand, it was poignant to see Harold John Frederick Gourley, who had designed the scheme for which, as head of Binnie's, he was also, in every sense of the word, the engineer, and also a man who had lost an arm and a leg in October, 1952, in the notorious telescopic train accident at Harrow and Wealdstone, go along at the rear of the queue, his stick hung in the waistband of his trousers, to shake hands with the royal personages who appeared quite oblivious to the part the man had played in the great works they were opening. I am happy to record in the next chapter, however, the signal honour accorded to Gourley by his peers in the Institution of Civil Engineers.

The queen and her spouse were too busy that day to attend the lunch in the vast marquee set up in the valley below the new dam to celebrate the occasion. The royals had several calls to make in the principality, and it was left to people like Councillor Percy Morris, J.P., M.P., the mayor of Swansea, to preside at the top table.

Shortly afterwards, in August 1955, the mayor opened the pipeline at the Bryn Gwyn Treatment Works and the new water was made available for the borough of Swansea with only a few days' storage left in the existing river Cray and river Iliw reservoirs, the then chief sources of supply.

I was now consumed by the desire to complete the settlement of the final account for the parts of the Usk scheme that had come under my purview. In the future, Swansea had nothing projected to compare in size with the Usk scheme, so I took the view that I had better look out for another job. This turned out to be as a resident engineer for Nigerian Railways. The appointment was to commence on January 1, 1956, giving me only a short time to agree to the final measure and account with friend Matthews, who was acting as agent for the contractors for this all-important period of the contract.

The intense heat of the tropics would be a great contrast after the severe winter in the U.K., so I persuaded my now opposite number to join me in keeping the office heating stoked up really high, with a view to acclimatising myself for Nigeria. This is a bit hard to credit, but there we were, in our shirtsleeves in tropical heat with the weather perishing cold outside the office, perspiring away on our papers before a cherry red tortoise stove. In this unreal atmosphere, we reduced ourselves and the differences between us in the final accounts, to find ourselves in virtual agreement by Christmas. Old Aylen, the C.R.E. at the reservoir, paid me a visit before I was about to leave the nearly deserted site office and compound, happy to see that we had no difficulty with the finalisation of the figures. The permanent establishment personnel for controlling the water supply had now taken up residence in the houses provided by the scheme on the site, close by the works now humming away merrily.

I could now take my leave of the corporation with a clear conscience in the knowledge that the work of providing the not inconsiderable extra water resources had been well done. Also, I had been elected as a member of the institution of Water Engineers, an A.M.I.W.E., in 1954.

Letters of mutual admiration passed between the consultants and the borough officials during which time I had a few days in the Cliff prior flying off to a far off land.

Once again, and in the preceding pages, I have been concentrating on the work aspect of my life. What had been happening to my private life? It was quite a testing time for the Prices, with me doing my best to pay 70 percent of my attention to work responsibilities and only 30 percent to my domestic ones instead of the other way round, if one is to ensure harmony in the latter area.

I did say that I was unhappy with my Singer Le Mans, which first served as a mobile office full of plans in the days when Kearsey and I wrote notes to each other and left them under a certain thorn bush in the woods. I changed it for the Rover fourteen horsepower 1938 model saloon, which glided along like silk despite its fourteen years of antiquity. It had a beautiful third gear engaged with no effort as one negotiated a hill. It also had a free wheel drive, which when in use gave me twenty-eight miles to the gallon, as against twenty-four with the ordinary transmission.

After travelling from Pembrey to Llandeilo in the morning, I used to leave the car at Jones and Griffiths's garage behind the post office. These two gentlemen gave conscientious service to our site Landrovers, which were kept under cover in their garage at nighttime. So 50 miles by private car and 70 miles on rough land around the site made a usual mileage for me of 120 per day, plus the executive control of a major contract. I found that after a time I had lost weight. The day came when I scaled at barely nine and a half stones, a drop of two stones in about two years.

It was in this period that, returning from work of an evening travelling through the inky blackness of the A 476 county road stretch from Llandeilo to Tumble, I used to have the feeling that my body was sometimes detached from my mind. This suggested to me that I had been building up too much nervous tension over a period of time, though I always fought like anything not to let anything show, trying to retain an outer composure as an example to the rest of the staff. My dinner would be shrivelled up with I got home about eight o'clock after an arduous day on the job, so I was not eating properly, having sandwiches in the day with half of shandy as I went round on inspection.

I resolved to go into digs, and after a short spell in a private hotel, I was fortunate indeed to find a homely place with Mrs. Evans and her family near the parish church. There I was able to make a conscious effort to restore my metabolism. It meant going to bed about nine o'clock with a cup of hot, sweet milk each night. As I lay under the covers I could feel the blood coursing back into my veins. With good food and rest in the evenings I could feel the return of my strength, so vital to keep on top of the work and win the private battle of mine. In parallel with my problem, Lin had her own difficulties with three children to look after down in Pembrey six days a week. Both of us were hard put to see each other's difficulties, being so fully occupied with our respective responsibilities. Before leaving for West Africa I do remember claiming her ear when we were taking a walk on the links, reflecting on what had been achieved for the benefit of the Swansea corporation.

"Lin," I said, "some of the tasks folk had to perform on site, like laying pipes, were pretty arduous to say the least. When Ernie and I kept vigil for twenty-six continuous hours at the Amman Valley Railway crossing, the men worked through while the rain fell continuously. It was a nightmare for Ernie and me. It was hell for the men doing the job, digging out, laying the pipes, jointing up, placing the concrete surrounds. Just one instance of six thousand, five hundred pipes, each and every one of them a story in itself. There are times when one can truthfully say that 'the part is greater than the whole.' This is when the task of the individual item calls for a greater individual effort of men's brawn and brain than the effort of the people who direct the whole show from on high. When we opened the Swansea Usk water scheme, believe me all could take just pride in their contribution in supplying the elixir of life of their fellowmen. For my part, I feel that with the Swansea work all behind me, I have now grown up as a civil engineer and I look forward with relish to that capital works job overseas."

"Let's hope it will lead to something when you come home again," was all Lin could manage to reply.

On the afternoon of December 31, 1955, I was travelling up to Heathrow. The countryside was covered with snow, and it was a perishing cold day. I had on my heavy topcoat and my heavyweight felt hat and I carried a brolly, the latter having been recommended in the literature from my future employers, the Nigerian Railway corporation.

I was going on a fifteen-month tour, another temporary job. As I was sitting in the plane on the runway, the engines were starting up in the late afternoon. It was to be a night flight—thirteen hours to Kano and then on to Lagos, the overall journey time being seventeen hours. It was my first flight in an aeroplane. Each of the four propellers was tried out in turn. As they were activated by the engines, flames shot out of the exhausts, the long wings seeming to vibrate excessively, almost flapping like the wings of a bird, and a piece of coating fabric was loose on the near side wing just in front of my near side porthole. Should I call the hostess? . . .

"Fasten your seat belts," came the words in lights on the bulkhead for'ard. I fastened mine.

We taxied down the runway. There was a sickening revving for a few minutes; then off we were as I sucked the sweet given to me by a hostess on the plane. The revving of the engines was intermittent as we gained more and more height. Suddenly the engines seemed to have stopped as the instruction came over "Unfasten your seat belts." We had levelled off and were on our way, the engines now being smooth and relatively quiet.

It was once again the end of a chapter for me and the beginning of a new one, which I will commence to relate by telling you about the rest of my journey to Africa. Cheers for now. . . .

13. Remodeled Railway

I was off on a flight to Nigeria. It was Saturday, the thirty-first of December 1955. It was my first flight in an aeroplane, and I had some misgivings. When I looked through the porthole on the port side. I must admit to a certain unease about the structural strength of the wings. *Not only do planes mimic birds by flying; they also flap their wings,* I mused. I was gradually becoming accustomed to the motion in the air, with the occasional easing of the throttle or whatever it was that caused the irregular drone of the engines sounding rather like the Gerry bombers who used to come into Liverpool during the war, with distinctive pulsing as their engines strained themselves with overloads of lethal high explosives.

We passed over the Channel in fading light, and we were on our way across France in the dark. After a while, I noticed that the inside engine on the port side had spluttered and the propeller was now at rest, so at that point we were running on three engines out of four. *That makes sense,* I thought. *The pilot must be resting one, with such a long trip ahead of him.* The engines had now assumed a different note. I wondered if I would ever get used to this flying. Hostesses flitted up and down the aisle, and we were treated to a nice dinner with special attention to detail, inclusive in the price of the ticket and served with that special, delightful charm air passengers were learning to take for granted in postwar aviation. There was a sort of silent camaraderie among the passengers travelling in this juggernaut of the skies.

After dinner came the crunch as someone on the starboard side mentioned that an engine had petered out on the other side of the plane. Now we were running on half-power, as the engine on the port side next to the cabin was still immobile. There was now a heavy drone from the two hardworking engines still going, and on came the bulkhead lighted notice "Fasten your Seat Belts," with an announcement from the loudspeaker: "We will be putting down in Tripoli to check on the engines."

The custom was for the flight to negotiate the Libyan desert in the hours of darkness, but now our timing was uncertain. When we landed at the airport we were quickly shepherded into buses. The repairs to the engines were expected to take twelve hours, and so we were going to stay the night in a hotel in the Libyan capital. While we were on the bus, security officials came to collect our passports with the explanation that they would be returned to us the next day when we boarded the plane. One man, a few seats in front of me, took a poor view of this, clutching his passport tightly and declaring, "I'm a British citizen, and I won't part with my passport to anyone."

His reason bore weight with his companion sitting on the same seat in the bus. "Same here," declared that one.

There was some attempt by the officials to reassure the resolute Britishers of their need to ensure security, but this was without avail. I suppose many of us admired their stand, but most of us were so relieved to be safe on terra firma and in any case, our departure was very much in the hands of our host country, which had not sought our presence, that we saw no point in demurring, "when in Libya . . ."

We were ushered into the chalet accommodations of the hotel all round a quandrangle, with the balmy Mediterranean subtropical air sweet on our faces. The night passed uneventfully and in the morning we fellow bus passengers boarded our own bus and were about to go when along came the patriotic pair, rather woebegone under escort, having spent the night in custody hanging onto their passports but possibly wondering whether their gesture had been worthwhile and whether they would be on the plane with the rest of us.

Travelling at seventeen thousand feet above the Libyan desert in the heat of the sun had the disadvantage of a bumpier ride than would usually have been the case in the cool of the night. The huge man-made bird did not have the art of the eagle in gliding smoothly with the currents of air. Man-made birds tend to go in a straight line, and with the need to travel "as the crow flies" (not really), we found ourselves falling like a stone from time to time or being lifted up with great acceleration as if we were in the elevator of a skyscraper.

Far below the clouds lay the bland desert, and for a long time I sat in the lounge in the tail end of the aircraft, imbibing long drinks just to while away the time while I studied the scenes presented by Mother Earth. There were no more incidents during the flight, but when we arrived at Kano Airport, we had to hang around while passengers left and boarded the aircraft. It was midnight Sunday when, at long last I struggled down to the tarmac of the Lagos airport, one of a mêlée of disgorging passengers. Most of the passengers were being met by transport while I stood waiting somewhat nonplussed, a figure of fun for the small boys still wide awake and dancing about me, as I had to don my heavy clothes, overcoat, and hat and carry my bag and umbrella just to retain my possessions. The night was saved when along came a car from the railways headquarters at Ebute Metta and a young engineer presented himself to chaperone me to a room at the rest house. The end of the journey arrived as the electric fan played on my naked body under a mosquito net. My heavy clobber lay on the floor with my bags, and with just a thin sheet over my tummy I lay through the night. If I thought

I had done sufficient preparation by way of acclimatisation for the tour ahead when in Bryngwyn Offices, Llandeilo, I was wrong! I had got the temperature right, the humidity wrong. All through the night my thick blood pumped sweat from my body like I had double pneumonia. Nigeria, here I come!

The next day, at the headquarters of the chief engineer, I met the acting C.E., Ian Gillespie, as Ronnie Bridgeman was on leave in U.K. How solid stands a Scot when abroad.

"You'll be doing the Capital Works, including the remodelling at Port Harcourt," he said. "You'll need to take over from Phil Rutter in PH, and you'll need to buy a car."

While being filled in on the railway setup, I was ringing around the four main suppliers of motorcars in Nigeria. The U.A.C. Company was the first to answer the phone, the other three being engaged when I rang them. "Yes, we have a Black Vauxhall Velox saloon in stock at Aba." I bought the car over the phone, thanks to the helpful scheme whereby the NRC advanced the money at 3 percent per annum interest.

At the airport outside Port Harcourt, to which I had flown in a Heron of Nigerian Airways, I could see two men talking together as I came into the small building attending to the formalities for all passengers. Phil Rutter, R.E., and his clerk of works, Dick Creelman, had travelled up to meet me from Port Harcourt. We drove down to the outskirts of the town in Phil's grey Morris Minor car, and in the afternoon I looked in at our offices, upstairs at the end of the two-storey railway terminal buildings.

Phil let it be known that he would be pleased to get away from PH. He was really a design engineer, and the subtleties of work in the field did not fall easily in place in his mind. Struggling to fit turnouts into place where the running shed was out of line in relation with the approach railway sidings, Phil had been preoccupied on the site while the paperwork had been piling up in the office, to the consternation of Fadipe, the chief clerk, with his tiny army of clerical staff.

We had to spare time to collect my new car in Aba forty miles from PH, as Phil had to run me up there, but it gave me an insight of the nature of the country; the long, flat stretches of straight through the "bush" was an immediate indication of how the Eastern Region was being opened up. Phil told me how an Italian contractor was making a big contribution to the road construction activities, shaping and forming the laterite soil of the ground into straights between the trees cleared for the purpose, the route being given a pronounced crown in the middle, with rain encouraged to flow sidewise to roadside channels, mostly to soak away in the pervious laterite subsoil. A centre strip of black running surface was contrived by means of several coats of runny bitumen from imported barrels of Colas and mixing this with sand as the layers were built up into a suitable thickness to take the traffic, mostly "mammy wagons" packed with people. Life was everywhere, with the native people trekking along their village paths where these crossed the route.

We returned from Aba, Phil in his car and me in mine, a 1956 model Velox, with its wider vision through the rear window compared with the 1955 model it replaced and silvered bumpers and trappings to freshen its black cellulose finish, an engine of some two and a quarter litres being adequate for anything I was likely to have to do in my tour of Nigeria. In a couple of days Phil was away and I was left with his bungalow. As I went round the railway property my first day without Phil, there was a clatter under the car like fans using their rattles at a soccer match. I was on the unmade access road to the railway Police Flats, part of our construction programme, and someone had left a piece of steel reinforcement fabric on the road, which had been hidden by the rainwater, which filled a deep potholed rut through which I had had to venture. Three prongs had pierced the wall of my rear offside tyre.

Fortunately, the tyre was one of the newfangled tubeless ones where all parts of the tyre remained under pressure, so leakage was slow after I had yanked out the offending steel prongs, one of them sticking in the palm of my hand in the process, so that I was able to get back across town to my bungalow with a loss of only four pounds per square inch of air pressure. There I had an opportunity of using the repair kit, with its rubber dowels pushed into the tyre using the bodkin tool and latex solution to make a perfect seal or in my case, three perfect seals, as I was to have no more trouble with punctures during my tour.

Back in the office, I was beset by a Bombay-type situation, only instead of Apathuri at Bombay, it was now Fadipe in Port Harcourt. Fadipe was so like Apathuri in stature I could not believe it; one African, one Indian, both genial, benevolent, lovable characters, family men of great local presence and ample, wobbly physical stature. But in addition, Fadipe was chief clerk, "finish." He had to answer to no European clerk. He had his own technique. Taking my first letter from the high stacks on my desk, I asked him what did so-and-so all mean. "Well, sir," he said, "if we say to them such and such, they will answer so and so and then we can reply in this manner and finish them." It was artistry. The man was a philosopher.

However, I soon had a problem. There was a lot of back correspondence, and Fadipe and his nineteen clerks were not going to be much use in disposing of it. What I needed was a shorthand typist. "It's not possible," explained Fadipe. "There are no shorthand typists in PH.

Mr. Restrick tried to have one all the time he was on seat, before Mr. Rutter, and there was no one to do such a job." Well, signals were flying in and out using the railway code, which kept the pot simmering, but contractual correspondence and fuller reports were heavy going. My evenings were being spent sifting through the files that I took home to my bungalow, the operation occupying me well into the night after retiring to my bed after a dinner, usually of curried meats, rice, and about seventeen side dishes. I did not require clothes, as I perspired under the mosquito net, reading the files into the small hours of the morning. After a couple of nights I began to get the picture of the necessities of the jobs in hand.

I was filling up my car with petrol during the first week I arrived when my Taid took action from above to help me with my difficulties. He spoke to me via the garage attendant. "Sah," he said, "I have a brother, just a boy of fifteen. He cannot catch job. He just leave school after learn shorthand."

"Send him to see me on Sunday morning," I suggested.

The boy arrived at my bungalow at seven o'clock. I began to work with him to see if he could be of some use.

To begin, I was like an Eskimo talking to the man in the moon. I was dictating too quickly in a strange English accent, the boy being brought up in a totally native way, with a pronounced Nigerian accent. I remembered the difficulty my own boys had had in Dunfermline school. I also remembered the time my older boy, Newton, had needed my coaching in sums to pass his eleven plus, beginning the scholarship year, taking one hour per sum but eventually getting through his home-work of eight sums in the expected time of forty minutes. So I went much more slowly, articulating very carefully, more intelligibly to him. Soon he was getting it back. By nightfall he had some replies to letters I had dictated safely in his notebook, and he had the ability to transcribe them. I asked him to report to the office the next morning. When I got to the office, Fadipe was beaming. "He is getting it," he said, delighted. "Nobody has been doing shorthand here before."

After that, the administration side became straightforward. The correspondence was up to date, there was the teleprinter for urgent matters, and the measurements for special instructions to be carried out by contractors or the direct labour gangs of the railway were always recorded in a cut and dried manner, with contractors' agreements obtained before the sun set each day. It was a system that worked and worked well.

What about the situation in the field? How were the works coming along? Two salient points emerged. The drainage was behind in relation to the expected rainy season, and the running sheds were out of position in relation to the approach sidings.

With regard to the drainage, it was essential to press the contractor, Costain (West Africa) Limited, to accelerate the laying of the Armco drainpipe network over the flat area. The turntable structure, a heavy annular foundation with side wall offering a dishlike catchment for rainwater, had to be drained into a soakaway because there was no reasonably convenient outlet into a drain, the land, with its laterite subsoil being so level for miles around. The soakaway obviously had to be suitable to take away rainwater from a tropical storm, so I had a word with the provincial engineer to see if he thought that it would be adequate to assume a rainfall of three inches of rain per hour, as we had allowed for our monsoon drains when I was civil engineer in Bombay. The P.E. thought I was overdesigning on this basis, but I did porosity tests of the subsoil and, went ahead with the soakaway on my own assumptions of rain. Thank heavens I did, because on the first day it rained we had five and a half inches with another three inches on the following day, so that the soakaway was just about the optimal size required to do the job.

Okpala was the Nigerian permanent way inspector in charge of the gangs laying tracks for the approach sidings to the running sheds, all part of the remodelling arrangements for the terminal railway station of PH. I had a quiet word with him about the fact that the sheds were not in the position they required to be to suit the curves on the drawings. Did he think he could lay the track in smooth curves from the crossing points to the sheds so the tracks would be serviceable for the rolling stock? He did. What is more, when he had finished his work you never saw better, sweeter-looking transitions, easier to check for acceptability once they had been laid than to calculate the modifications to the curves in the first place and then set out with a theodolite, thus saving valuable time for this expatriate engineer.

When the rains came, they were tropical! If I parked my car two feet away from the bungalow door, I would be soaked to the skin just stepping across those two feet from the car into the house.

My clerk of works, Dick Creelman, had to leave PH about the middle of 1956, and Arthur Senior, a Yorkshire man, came to take over his job. Arthur was an immediate hit with the local labour. The way he coaxed them to put together the new coal conveyor machinery for the coal bunkers at the docks was well worth beholding. No drawings existed. It was just a collection of parts sent out from the U.K., yet it went together like a child's jigsaw puzzle, thanks to Arthur's leadership and the innate intelligence of the Nigerians themselves. In the end the conveyor worked a treat and offered the mechanical means of handling the coal from the bunkers above it, which had just been built under the supervision of Peter Hilton, as R.E. for the consultants, Messrs. Coode and Partners. Peter and Pat, his wife, resided in the bungalow next to me.

Soon after Arthur came to PH, Fadipe had to go into the Roman Catholic hospital at Uyo for an operation for elephantiasis. He had been suffering with the complaint for years, but with the elapse of time, the swelling had become the size of bowling woods, with consequent unacceptable discomfort. So the clerical office was handed over by Fadipe, with great reluctance, to his assistant, Bamijoko, a slight, alert man with a cheery expression ever pervading his physiognomy. Arthur and I went up to Uyo to see the old clerk just before the operation and found him in good spirits in the superlative comfort of the hospital. Later we paid another visit, and when he returned to the office after his three months' absence he was really a new man, quite spry after the removal of those weighty appendages in the lower regions.

The Nigerian Ports Authority had decided to extend the wharf at PH, and the day come when Wheatley, Coode's chief resident engineer, made his number, asking me if I had any plans that might be useful for the site they had to exploit for the new works. As it happened, I had just the very thing. During the earlier part of my tour, Dick Creelman and I had set out a long baseline on the railway property, and from this work, which had been done in the cooler part of the day, from four to seven in the late afternoon, a survey for exchange sidings covering a vast area of bush site had been developed. Surveying in the heat of the day was too severe a chore for us oldish codgers, but I was able to enlist the help of a young messenger in the office. He quickly grasped the mysteries of turning right angles from our long baseline with its pegs at hundred-foot intervals, at the same time getting slasher gangs to clear his lines through the bush. The upshot was that the land had been pegged out, with surface levels ascertained, sufficiently for a start to be made by my newcomer friends from the Port's Authority.

It is interesting to remember that the day would come when the help I was able to accord to Wheatley and his assistant, J. L. Horsburgh, would one day be repaid to me in similar vein, when Horsburgh would be an incumbent at Tilbury and I an incoming R.E. to do a new job for consultants at Tilbury Docks, as I will relate in chapter 17.

Visits by senior liaison engineers from Ebute Metta headquarters occurred from time to time, when they generally came in inspection coaches. Also from Enugu, two hundred miles to the north, came an electrical engineer, Harry Deiton, with whom I was to form a long-lasting friendship. Harry took care of matters electrical at Enugu and PH.

It made a nice break for me to spend a few days with Harry and his wife, Mary, at Enugu. Mary was just about my match at tennis, my

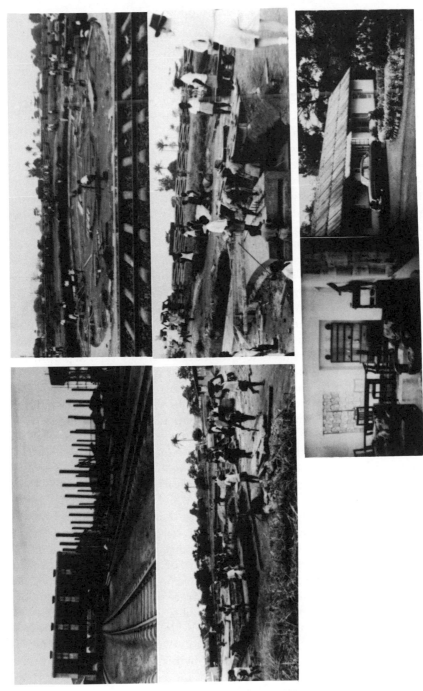

Port Harcourt, Nigeria, 1956.

last endeavours at the pastime being at One Tree Hill in Singapore ten years previously. The journey from PH to Enugu by car took me four hours, the way being mostly on straight lengths of flat, level road, though as one neared Enugu the terrain was hillier, with Enugu lying in a picturesque hollow down below the traveller arriving from the south.

At PH visibility was very limited because of the high rubber trees that covered the land in thick abundance in the area of the Bonny River, forty miles upstream from the coastline. As I was senior railway official, it fell to me to attend regular coordination meetings held by the resident and a planning meeting chaired by the district officer. These meetings served to ensure the smooth working of the developing port of PH. With trade growing, expansion of facilities for shipping, railways, roads, and town planning to regulate new building, it was imperative that order should exist among the various bodies serving the growing community in a changing civilisation.

Summary justice was meted out by the railway police to trespassers and thieves caught on railway property, where a number of squatters had the habit of carrying on their affairs in a traditional bush free for all. It was customary to see these people queue up at the police offices to receive their fines for transgression. This all came to an end when we finally established the boundary of the extensive railway land with a new security fence, several miles in length. There was a short, sharp upheaval as the bush hutments' occupants were evicted by the authorities.

Halfway through my tour in Nigeria, my family were able to join me for a period of two months while the children were on school holiday. Lieutenant Colonel Emerson, the Nigerian Railway chairman, and R. Bridgeman both made "on line" visits during the time we were all together as a family, and the weeks slipped by all too quickly as we managed to make visits to Aba, Enugu, Ikot Ekpene, and the River Cross separating the Efik and French Cameroons territories. The visit by the family broke up my tour very nicely.

It was while I was in Port Harcourt that I was accepted by the Institution of Civil Engineers as a senior corporate member with the standing of member of the institution. In those days, corporate members were either associate members, A.M.I.C.E., or members M.I.C.E. These gradings were renominated on March 19, 1968, to members and fellows respectively. Transfer to the senior corporate was contingent upon having had a minimum period in responsibility for major works and, of course, in taking the trouble to be sponsored by one's peers and applying for the transfer. In the chapter "Drops of Water," we refer to Harold John Frederick Gourley's signal honour by his fellows at the institution. Well, it was to my delight that my certificate dated November 20, 1956, for the transfer to the senior corporate membership was certified by H.

J. F. Gourley in his capacity as president of the Institution of Civil Engineers for that year. The fact that H.J.F.G.'s signature was made with his remaining left hand is always a reminder to me of the courage and character of the man who continued with his job after such a horrendous experience in the train accident. In the same year, Colonel Emerson, chairman of the Nigerian Railways Corporation, was accepted as an honorary member of the institution, so I felt I was in good company.

When it was known that my tour would see the completion of the current programme of capital works in PH, I was asked to visit Jebba, a station in the Western Region, low down alongside the river Niger, some four hundred miles north-west of Port Harcourt, which itself lies in the Niger delta, the capital of the Rivers Province, forty miles from the Atlantic Ocean. Jebba offered the prospect of work had I wished to return to Nigeria on another tour. The whole station had to be raised several feet to overcome the need for an extra shunting engine, which was required every time a train had to pull out of the low-lying station, as one engine was not powerful enough to get a train away up the hill to the north.

Eventually the tour came to an end at the end of March 1957, and with it, the purpose of the tour, to widen my experience and save up some money towards a house of my own, which I have to say had escaped me during my time at Swansea, which had offered me experience only, no savings to speak of.

Colin Gosling, a pupil engineer who had been doing a good job of work in the drawing office at PH, since he had arrived some months previously, was due to go to Lagos on transfer, so it was arranged that we should travel to Lagos together, as we each had to take our loads in our respective cars. When a few people have achieved something together far from home, it is hard to have to part, and it was with some sadness that we took leave of our friends at PH.

On the trip to Lagos Colin and I were to be glad of each other's company on the two-day journey by road, the incidents of the first day I shall recount in the form of a short story titled: First Stop—Benin.

By eight that morning I was leaving PH with pride and relief, the fifteen-month tour with Nigerian Railways over. It was April 1957. The new works would remain as a fitting railhead for the Eastern Region. But PH was a hot, steamy place.

My inside mirror revealed the reflection of a dark grey consul car through the wide rear window of my Velox, the consul emerging from a gateway and following me like a speed cop. "Colin's a dependable chap," I reflected.

The two cars accelerated away to travel in convoy, on the only road to the north. My steering felt lighter than usual. I thought of my loads in the boot and on the rear seat of the car. *Take it easy or something could happen,* said my conscience. Our destination was Lagos, the Western Region capital, 550 miles and two days' journey away. Nigeria was developing quickly but still young in terms of "Western standards." I had to embark on the M/V *Accra,* which was due to sail in a couple of days, while my young engineer was being transferred to complete his Nigerian tour at the Ebute Metta railway headquarters at Lagos.

Nine o'clock and some fifty miles on saw us already through the town of Aba, cruising on towards Owerri, keeping a half-furlong apart, doing sixty miles an hour, and holding the crown of a long, straight stretch of road. Much of the route was the same, a narrow black running carpet flanked on either side by rolled red laterite soil verges, sloping slightly sidewise to open channel drainage ditches at the edge of the open corridor cut through the trees—a regular swathe through the bush.

Occasionally we would meet a mammy wagon. It usually took the form of an old lorry with wooden bench seats on its flat platform body, conveying a load of the populace with their paraphernalia. The driver's cab was chunky, square, and home-made. A biblical epithet emblazoned across the front would exhort the traveller to "Prepare to meet Thy God," or proclaim, "Thy will be done." *The Bible is overtaking the witch doctor,* I mused.

On the straight, level lengths an approaching wagon would zoom in from the size of a dot to that of a bus right in one's path, and a battle of wills would then ensue. Two seconds away, it would veer deftly off the black surface onto the verge and zip skilfully alongside with perhaps a yard of clearance as my car held its line on the road. This was a sort of bush highway code and worked infallibly—almost—and to prove it there were but few wood crosses or images or vehicles remains decorating the fringes of the woodland where the vehicles had "gone for bush" and made an error or judgement.

After about 150 miles heading northwards, we slowed down as we entered Onitsha, gentling down a ramp leading from the road to board the waiting Niger River ferryboat. I vaguely remember it as an old pontoon-cum-barge affair, already full to nearly sinking. Mammy wagons, bullock carts, bicycles of the more posh type, foods of all kinds, live chickens, plantains, yams, fresh and stock fish, women with babes on their backs, old men, young men, and children dressed in remnants. More people were packing into this "Noah's Ark" affair carrying their property "put for head."

"Come, come," chorused a knot of strong young men watching the approach of us two "English Europeans," quickly parting to make a passage, and half-guilding, half-lifting, our cars from the gangway ramp, to allow us to squeeze into the seething mass at the stern of the vessel.

As we chug-chugged heavily over the Niger, the brown human throng chattered noisily, the little ones swarming round our cars merrily, their white teeth laughing incessantly while we munched our packed lunches and accepted a banana or two.

At the western side of the river was Asaba. Those last on were first off, as the craft came in stern first, and we were off on the second leg of our day's run to Benin, halfway between PH and Lagos. "The worst is over," we had agreed as we left the river, but we were wrong!

When it suddenly cooled, I quickly wound up the windows, but not before the first splash of rain wet my shirt. Distant bright flashes showed we were in for an electric storm.

"Typical, tropical, and torrential," I mouthed as the wipers, working frantically, failed to cope. My hand, on the column lever, knocked out the wiper blades to give nature a chance. Daunted by the prospect of pressing on farther, I was very relieved to see that the sheer weight of the rainwater was cleansing the sloping screen curving before me like a big bubble. Adjusting my speed to the watery view, I felt like a fish in a tank, staring outside. Time was passing slowly, but we had passed the shadowy shapes of the town of Agbor. Eyes sore with concentration, I was beginning to weigh up the choices, to carry on or to stop. Taking a sweet from my breast pocket, I popped it into my mouth, sucked hard, studying the misty red quagmires flashing by for a place to come to rest. I was overruled by a second opinion. *Don't stop. Keep at it,* said my mentor within me. It was foolish to stop in the bush, where witch doctors' voodoo magic still prevailed against Bibles and influenced elders of the villages in meting out summary justice on the unwary stranger.

Suddenly there was a brilliant flash and thunder crash and little brown bodies were leaping and running in all directions as I felt a bump-bump on the road. Easing up the car, now rumbling over a rougher stretch, I passed through a small village settlement on either side of the road, with the villagers washed out of site in their huts apart from a few little ones, topless and bottomless, splashing about in the open-air shower from the skies.

Then the rain stopped. There were more huts, and a skyline of high trees heralded our approach to the ancient city of Benin, capital of the timber province of the same name.

Standing in the rest house car park, I looked at the front of the car, its bright silvered bumper bent inwards about a third of the way from the near side, with what seemed like a piece of wood wedged behind and below it. The object was spongy as I pushed it down to drop on the ground. Picking it up, I felt a shiver run through me.

"How did that happen?" My companion had joined me and seen

the dent in the bumper and what I held in my hand—small brown foot, washed clean of blood by the rain.

I simply shook my head, perplexed.

Then he said, "I thought I saw two little kids, flat on the verge. They looked to me as though they had been there ages."

With a silent prayer, my conscience took hold of my tongue. "It might have been worse, Colin," I uttered, in a whisper. "They might have been . . . children." I had hit goat kids!

After that, the next day's journey through the university town of Ibadan to the railway headquarters at Ebute Metta was uneventful, and to recall the period would be just an anticlimax. At the Lagos end, tickets had to be arranged for myself and my car, and eventually, I found myself looking back on the harbour at Lagos as the ship pulled out from Apapa Wharf, the ship being the new *Accra,* the old one which I had travelled on as a student engineer when at Liverpool University (see chapter 4) having been sunk by enemy submarine and aircraft three hundred miles off Ireland on July 26, 1940.

Jack Parry was an electrical engineer whom I met on the ship. We were soon chatting away about our time spent in Nigeria. Jack stood at the ship's rail on the promenade deck, taking pictures using a splendid-looking camera with three lenses, another camera being slung over his shoulder, Japanese style.

As the harbour fell away in the distance, on April 9, 1957, the thought of going home rested sweetly on my mind. Jack and I had the measure of each other by the time we went to our respective cabins. He hailed from North Wales, but in the twelve years since the war he had built up a business near Nottingham, in a factory manufacturing industrial electrical equipment. He was now clearly well established, and his visit abroad had been to advise on underwater harbour lighting. It was to be a happy boat acquaintanceship. Jack was an extrovert who loved to be in the centre of things happening among the passengers. Soon after we were out on the high seas, with ship's games and the customary sweep each day guessing the ship's mileage, Jack threw a morning party, inviting notables among the passengers, dignitaries, and officials returning home. A good time was had by all at the morning cocktail get-together, yet the bill at ship's prices was so modest, with I suspect, ship's officers having their drinks included on their bar accounts, that Jack felt it quite a letdown to be debited with a trifling sum. "Bloody silly," I seem to remember as being his reaction to that situation. On the other hand, the company had been mixed and friendly and it had been an occasion to set in motion new acquaintanceships with Jack the instigator. Governor General Sir James Robertson, G.C.V.O., K.C.M.G.,

K.B.E., wished me a "cracking job" when I returned to the U.K., and his words have stayed with me long.

As I was seated alone at my table in the dining saloon, over came David Aitken, returning home from service with the Nigerian Ports Authority at Lagos. "I see you are travelling alone," said Dave, "Our younger child is taking her meals in the cabin, so why not come and join us?" They were three at a table for four, Dave, his wife, Connie, and the daughter, Judith. I was very pleased to have their company at meals and at other times during the rest of the trip back to Liverpool. At Las Palmas everyone went ashore. At the Catalina Hotel flower-selling maidens pressed their flowers on us as we made to go in the taxis, their faces lit up like angels in the sunlight.

Back at last at Prince's landing stage, on April 22, 1957, we had made our last leg of the journey through the choppy waters of the Bay of Biscay up through the Irish Sea and rounded North Wales to cross the Liverpool bar at the mouth of the Mersey. Soon there was hustle and bustle and I lost touch with Jack and Dave and family as they went about their own arrangements. My parents had waited to meet me off the ship. It did not take long for my car to be offloaded from the ship, and there was little time wasted in reregistering the car with a Liverpool number. I was able to spend a night in the old home at Ewloe Green.

It seemed to take an age driving down through Wales the next day, but finally Lin and the children and I were united after the months we had been apart, since the school vacation. There was now a substantial fifteen weeks of terminal leave to wind up my contract with the Nigerian Railways Corporation.

Another chapter in my life had drawn to a close. . . .

14. The Great Wide Way

The "cracking job" wished on me by Governor General Robertson, fellow passenger on the second edition of the Elder Dempster vessel M/V *Accra* as it came from Nigeria, was to come my way. My leave of fifteen weeks following the tour in Nigeria would give me to the end of July 1957 to scout around for another post, likely to be as it had been up to now a temporary situation, as is the case with so many men following my itinerant profession.

After spending the night in my parents' home, I wasted no time on the twenty-third of April, in journeying down to my wife and three children at home in the Cliff, Pembrey. It was nice to be back in the U.K., and with my last year's model Vauxhall with its Liverpool registration number WKD 937 and only eleven thousand miles on the odometer, we were able to enjoy running about touring South Wales in the pleasant months of the year.

One day there appeared an advert in the *Daily Telegraph* for civil engineers. It was a modest entry offering experience in earthworks, drainage, and concrete. I drew up my two decades of experience in tabular form so that it could be seen at a glance and sent it off, receiving a reply in the course of a few posts. In the London office of Sir Owen Williams and Partners I appeared for interview with the man himself seated at his desk, a younger man seated on his left side. As they rose to greet me, Sir Owen made himself known and introduced the younger partner as O.T.

"Mr. Owen Tudor Williams?" I queried, having checked up before I came that there was such a name in the list of members published by the Institution of Civil Engineers.

"That's right," said O.T.

"We do roads," said Sir Owen, which was news to me. He was not a man to exaggerate. "I see you're a *resident* engineer."

"Quite so," I replied. "If you have to give me a title. That's what I have been in my last few jobs."

"All this," he said, looking at my table of information he had out on the desk. He and O.T. went through my previous experience, seeking sufficient elucidation to meet their requirements.

I waited outside the room for a few minutes, talking to a pleasant small-statured man working on a drawing board. When asked to return,

Sir Owen said, "We can offer you work as an R.E." Then he made me an offer, fair in the circumstances that I was pleased to accept. They were the consulting engineers for the M1, the "London to Yorkshire Motorway," as it was to be known.

"Of course, the job is only temporary but there should be about three years' work," said Sr. O———. I said I was quite pleased with that.

O.T. added, "If a man can see himself fixed up for three years, it's not bad." I would come to know that "Not bad" was an expression they frequently used, their low-key way of describing favourable circumstances.

I went home delighted in the knowledge I would start at the beginning of August in a job on the M1 at a field office set up in Northamptonshire. When I told Lin, she had mixed feelings. She had lived all her life in South Wales and was uneasy deep down at the idea of leaving the Cliff, Pembrey. When I went to look for a house in Northampton on the following Wednesday, she made it plain I should be back on Friday or else!

Off I went to Northampton to stay two nights in a commercial hotel and see the staff at Welton, the firm's offices there being busily engaged on the design of general details for the road, drainage, alignments of side roads, and innumerable plans, land negotiations all being done on a noncompulsory order basis. The firm's task for the sections of route in the planning stage would involve the production of thousands of plans—no less than fifteen thousand were turned out for the fifty-five-mile length from Luton to Dunchurch alone—many of them having to be coloured. Drawings for the 138 bridges for that section, as well as for the bridges in Hertfordshire on the length being engineered by the county engineer, Lieutenant Colonel Foliatt, were being produced in the London offices of Sir Owen Williams.

A quick introduction to the chaps at the Welton office and Robert Denton Williams, nephew of Sir Owen, R.D. to everyone, administrative officer at that office, introduced me to a number of estate agents in the area and then left me to try out my luck. Returning home on the Friday, as I had promised Lin, I had details of two particular houses that seemed to fit our needs, but down in Pembrey we plumped for the wrong one. At least, as is often the case, we were to feel we had missed a chalk in not electing for the more commodious house with its study and its one flat acre of land. I must admit to an error of judgement on the cautious side in taking a smaller property, this being my choice in view of the uncertainty of the future in a temporary post. I suppose when a man has scraped pound by pound for twenty years, he is not likely to be given to throwing away the money he has saved.

In July, Lin sold the house in Pembrey. When we arrived at our modern, smallish detached house on the Welford Road at Northampton, with its prefabricated garage, the problem of squeezing in the furniture of a two-hundred-year-old cottage property, including antique pieces and a heavy Eungblut piano bought by the old lady in the days of the First World War, was like getting a quart into a pint pot. Lin, homesick for Pembrey, looked at the way her furniture was not right for the house, and vice versa, and kept complaining to the removal men that the house was "like a box." It would have been inexpedient of me to remind her at that point that she had thought of the Cliff latterly as an "old barn," in a weaker moment, when doing her house chores, in the old place. I have to admit, with hindsight, I made an idiotic choice, but I did not weigh up the situation properly, being consumed with a desire to live within our means and not to draw on the proceeds that Lin had for the cottage, working on a principle I have retained through life.

To ease the wrench from the seaside we had all been used to in South Wales, the following day, Sunday, we decided to take a run out in the car as far as the Wash. Well, a bank we climbed up to see the sea gave us a view of the Wash, a vast muddy marsh, which pleased us not a lot, but one little item did. When we went into the town of King's Lynn, I was looking into a tailor's shop window with a Harris tweed jacket displayed in the centre priced at £6 : 10s : 0d. "Just the thing for the site," I said to myself, but it being Sunday, the shop was not open. At that moment, along came the manager to pay a call in the shop, and I persuaded him to sell me the coat, which I used for nearly thirty years, albeit with a second lining.

Life in the Midlands was a rich experience. We were soon settled in as a family and each of us had a new direction to follow, yet we were all together in our own home and, as it was to turn out, for just a few years, after which our teenage children would fly away from us to make their own way in the world. On the first Monday in the new place, Christine, our ten-year-old daughter, was lonely, missing her friends in Pembrey, but the following day she met a neighbour her own age.

On that Wednesday in came Christine. "I've got ten friends, Mummy," she said with delight. She was the first of us to feel at home, though when she attended the village school, at Chapel Brampton, she was to experience the almost inevitable setback of changing schools. Her scholarship was adversely affected so that she was destined to miss out on going to a grammar school, such as she had fully expected to attend had we remained South Wales.

Our two sons reported to the boys' grammar school. Their progress was also affected, the new school of nine hundred boys being larger than the school of Llanelly. There is never a good time to change schools in this way. The children had moved during the summer holidays, which lessened the effect of the change. After a short spell in the Duston Secondary Modern, Christine was admitted as a day student of Overstone School, of which the then headmistress was Miss Charlesworth, a kindly, homely, gracious lady.

Lin and I and our whole family were soon welcomed into the parish of Kingsthorpe by Donald Andrews, the rural dean, Kingsthorpe being in the see of Peterborough, with its beautiful cathedral, heated with the old slow-burning, French combustion stoves. I gathered that the vicar, Canon Andrews, had also been a civil engineer in his younger days. At all events, everyone was made welcome in the parish of Kingsthorpe, and Lin soon found friends in the village.

On the work front, Lt. Col. Bill Tyrrell and I looked at the plans and particulars of the proposed new M1 motorway, to be known as the "London to Yorkshire motorway"—a great wide way, which would ease the traffic on roads to the north, the A1, the A45, and the A6. I was given to making copious notes and making numerous sketches to comprehend the detail of the endless plans that had been prepared. Tyrrell and I spent weeks looking at the route, walking over it, step by step, examining salient points, and familiarising ourselves with the country and the farms affected by the plans. On one occasion, I fell into a ditch up to my armpits in mud, and Bill Tyrell pulled me out.

The day came when I felt I had made enough notes and sketches for ready reference for the motorway to have been built from them even if all the plans had been lost. The day also came when Sir Owen arranged for a party of his senior engineers, designers, and prospective residential staff to visit the work in progress at the site of the new Preston Bypass. We were received by James Drake (later Sir James) and his county engineering staff on site supervising the works, who conducted us around the accessible points to witness the absorbing activities proceeding, with mammoth earthworks in ground that called for vast open cuttings with flat slopes and considerable deep excavation below formation levels to enable backfilling of new materials strong enough to support a heavily trafficked road. The visit served to illustrate the type of operations we could be carrying out in a few short months down in the Midlands, but our operations would be on a far greater scale, initially, in terms of length.

As we gazed down into one great cutting, with machines crawling all over the slopes of silty clay and hardcore being tipped to fill a great void below formation level way down below us, I ventured to observe, "It can be done." Sir O—— nodded. It was a timely visit. We were under no illusions.

It was not so very long after this visit that tenders were considered for the four contract lengths of the M1, and in the result, the four contiguous lengths stretching from Pepperstock junction, south of Luton, in Bedfordshire, to Crick, in Northamptonshire, with a spur off to Dunchurch, near Rugby, were let as one contract to the contractor presenting the lowest tenders, the civil engineering contractors, John Laing and Sons Limited. That mammoth contract for fifty-five miles of new motorway to be built in a period of nineteen months was probably, and probably remains, the heaviest contract of its kind ever let anywhere as a single contract, the rate of construction expected making it a quite phenomenal task in the decisionmaking, the planning, and the negotiating. It is to the credit of our country that there existed the necessary acumen, verve, foolhardiness, and enthusiasm, to strain every nerve from the day of the inauguration of the works by Harold Watkinson, the then minister of transport, to the opening of the M1, seventy miles including the sixteen miles tackled in the same period by other contractors for Hertfordshire, south of Luton. The opening took place on November 2, 1959, with Ernest Marples, the then minister of transport, to preside at the ceremony, calling the M1 "a powerful weapon" and Sir Owen saying here was a road that would be seen "for all time," a profound pair of comments. The provision of the M1 at an overall construction cost of some £350,000 per mile, supervised by an official site staff of numerical strength a fraction of that employed in subsequent years, was invaluable as the forerunner of our motorway system. The M1 was a great proving ground.

Before tenders had been examined, it had been indicated that Tyrrell and I would each supervise two contracts, each with the title Senior Resident Engineer, but when it was known that the four lengths would be amalgamated into the contract, it fell to me to be site representative for the engineer, with the nomination Chief Resident Engineer, with four resident engineers assisting me, covering the four lengths of motorway as they had been separately itemised and priced, with individual rates applying for similar operations in the four contract sections, A, B, C, and D, the D contract being at the northern end length, sixteen miles initially, but eighteen miles finally, and supervised by our friend Bill Tyrrell as resident engineer. Arthur Price, Vic. Poulton, and Lynn Roberts were our resident engineers on the sections A, B, and C, respectively. Sadly, both Bill and Vic are now deceased.

We resident-engineer types each had opposite numbers operating for the contractors, namely, John Michie, the chief project manager, and from south to north project managers James Cryer, Bob Spencer, Ron Smith, and Dougy (D.S.) Elbourne.

My deputies at our central site control offices at Newport Pagnell, Sir Robert Marriott, Kt., and W. D. R. Kerr, paired off nicely with John

Michie's deputies, John A. Pymont and John Gregg, Jr., in respect to their duties on planning/progress and finance, respectively. Alistair Foot, formerly subeditor for the *Northampton Chronicle and Echo* and later of achievement I shall touch upon in a later chapter, was my diligent office administrator, unceasing in his enthusiasm in getting the paperwork through with the strict timetable precision demanded of the administration.

The contractors were carrying out a task of construction that would have taken half a million men to do in the same time without mechanical plant, but managed with a peak labour force of over five thousand, with the aid of two and a half thousand items of plant worth £5 million, at that time.

Official visitors to the site of the works numbered more than the workforce at its peak, about ten thousand altogether. It must have been the most inspected job ever done so far as outsiders are concerned, yet though many problems were solved with the advice of external specialist technical officials, once any contract of this nature is started on site, there is little that can be done to upend the original parameters of the documents, determined in turn from money allotted standards decided upon by government department, design codes, and what have you. The standard M.O.T. pink specification that applied at the time, the road notes of the Department of Scientific and Industrial Research, later called the Road Research Laboratory, and codes of practice all had to have later revisions in the light of the experience being gained and the level of road usage changing in process of time. It is not uncommon for any developer to be sparing with funds for such as site investigations at the beginning of a venture. In a world where *cash flow* is the watchword, it is small wonder that what is taken to be a practical economic solution has more stress placed on the economic than the practical, sound, certain solution. In this government departments, answerable to the Treasury, are only human though they are faceless when the world pontificates with the benefit of hindsight, the finger being pointed at those not having the protection of a department. Thus it was that when the new phenomenon was opened to traffic on the second of November 1959 the good old British "open road" gave license for the motorway to be used as a racetrack, with one person crashing almost immediately, in a spin at about 140 miles an hour, could not be blamed on any official, nor could the fact that the shoulders could not withstand the number of heavies with overheated engines assailing them on a broad front at a time when the surfaces had not had the benefit of having such traffic using them during summer months. Like I said, the M1 was a proving ground, Preston Bypass having just pipped it to the post in proving the need for more positive drainage than had been allowed by the boffins.

During the twelve-month period of maintenance, in which any design or construction faults had to be corrected while the county council highway departments attended to routine upkeep, repairs to damaged guard rails, salt applications, and fair wear and tear, all concerned with the design conceptions, maintenance, and police control of the traffic, learned the lessons that would pave the way towards future motorways in this country. At the same time at Newport Pagnell offices, my staff, now down to a dozen, was busy with the contractors quantity surveyors' staff in settling the final accounts, with all the additional items established inevitably on a pioneering works of such a size. I was lifted by the fact that my engineering staff, so often not given to a preponderance of measurement duties, stuck to their tasks in working on the figures without complaint, so that the multitude of detail was dispensed with creditable efficiency.

At the end of the maintenance year, the handful of staff still associated with M1 duties were accommodated in the firm's Welton office compound, to see the measurement through to the end and to supervise the strengthening of hard shoulders, as a result of the experience during use. Many who had been on site during the construction of the fifty-five-mile stretch of M1 completed were now looking forward to the extension of the motorway up to Doncaster, where it was to join up with the Doncaster Bypass.

It had been hoped that continuity of the M1 up to Yorkshire would be possible when the first contracts, A, B, C, and D, had been completed, but the great wide way to the north was being held up at Crick in Northants. Farther north in Leicestershire, in the area of the Charnwood Forest, protesters had held up the great push northwards at the design stage. The result was that instead of being employed to supervise work actually expected to be proceeding on site up to Yorkshire, staff had to be diverted to other work. A few remained footling about on measurements and claim shooting, others were filling in their time switching the priority to the Midlands link with the M5, Birmingham, M6 routes to the north on the more westerly side of the country.

Now, the right to protest against intrusion is our sovereign heritage, but while we sympathised with the protestors, it did keep some of us gnashing at the bit, so to speak. I had spent hours of my own time weekends walking and driving up the whole route planned for the M1 from Crick to Doncaster and had the vision of the work being done in four contract lengths, each of about the same order of size as the completed contract D with our experienced people from the southern end leading the organisation required. But in the time it took to deal with the Charnwood problem, the staff had dispersed, even if the people with the say so had had the same conception as myself. As it came about, there was a rethinking of the contract lengths by the Ministry, enabling more contractors to participate in the action, and a reorganisation by the consultants, who needed new executive partners to deal with more liaison personnel at the Ministry. My own position was changed. The salary payable, now laid down by the Ministry, was actually reduced somewhat, which did not worry me too much, though it seemed a strange way to recognise the service of the men in site charge, now to supervise fourteen contracts instead of four and eighty-odd miles instead of fifty-five. I was uneasy that the direct personnel contact with the firm's partners was suddenly missing.

For the first nine months after construction on the M1 extension began on site, when substantial progress was made on earthworks, piling, and bridge foundations for the first four contracts—E, F, G, and H, some twenty-eight miles of the project—I set about the task of the man leading the organisation in the field, with five immediate assistants at the Control Headquarters at Leicester Forest East, where there was also established a most helpful partner of the firm of Quantity Surveyors, George Corduroy & Ptrs., Frank Roe, appointed to attend to the certificate work for all the contracts. So here was specialist help for the measurement coordination, so advisable for civil engineering schemes of mammoth dimensions, so that civil engineers can devote their time more to civil engineering problems. In addition to this, our own staff was already nearly three times the number we had on site for the fifty-five miles in the south, completed in nineteen months as against longer periods in excess of two years each envisaged for the works to progress northwards over a much more protracted time scale. Experience in the building of later motorways has led me to observe that the new level of supervision set on the M1 extension was appropriate for the contracts in hand.

As always, every effort to help the work was accorded to us by the various authorities concerned, the county surveyor and the chief constable of Leicestershire being especially cooperative and a valuable link with the community.

However, the time came when I could see before me too many years with more of the same, contracts being let, six or so miles at a time, till we got to Doncaster, and I felt that this, coupled with a less personal contact with those who mattered in the firm as such, placed me in a position I had not previously known in quite the same way. Eventually I gave in my notice, saw O.T. in London and, with tears in my eyes, not my usual stiff-upper-lipped self, gave up the job I had set my heart upon and identified with more than any other. The moral of this is "Don't count your chickens . . ." The final word came from the old man, Sir Owen, in a letter. "It was inevitable," he allowed. The man I held in high regard had struck the spot—I had become a square peg in a round hole.

Now so much for reflections on my working contribution to that great wide way the M1 motorway. It had certainly been important to me, but a man's work is not the only aspect that matters in his life. There were personal relationships and happenings to mention before we move along to the next chapter of my journey through life, following the relinquishing of my post on M1 in April 1963.

My family had stuck it out in our small house in the Welford Road, Northampton, until the beginning of 1961. Somehow it seemed important that a roomier place was available by the time my elder boy, Newton, attained his twenty-first birthday. On February 23, we combined his birthday occasion with a housewarming party at our new home, a five-bedroom house in Boughton Green Road. The party for young people went well, and we were soon settled in with no thought of having to leave for the foreseeable future. The house had a large garden, a lawn, and a small spinney of newly planted fir trees, the whole amounting to an acre and a quarter—ideal to establish as a family homestead.

High poplars growing at close intervals in a fourteen-foot-high mixed hedge marked the quarter-mile-long boundary around the property, casting shadows on one or two complaining neighbours residing over the boundary, with its seven frontages. Thinking about it, two complainants out of seven neighbours was "not bad," as the law of averages goes. The boys and I spent time reducing the hedge to a less unwieldy height of eight feet and did surgery on the poplars, both to give some relief lightwise and to provide fuel for the underfloor-draughted Bell fireplace in the large lounge.

One day, my second son, David, and I were having a go at a tree in the front with Mr. Pearce, who did some gardening for me. It was Saturday and drawing on midday, and friend Pearce was due to go on duty on the railway. "I'll have to go now," said he, making to hand me one of the ropes we each held in our hand to hold the top of the tree being sawn off by Dave, up the high ladder. The trees were wobbling, with a wind blowing towards the main road. It was the last tree we had to top, and thanks to my son's sense of duty in sawing away at an accelerated rate and my beady eye persuading the gardener to give us another ten minutes of his time, or the railway's, the top was severed and safely lowered, with the three of us feeling very relieved.

In July 1961, Lin, Christine, and I took a holiday, having decided to go to Yugoslavia on a coach trip, the tour starting from London, with a coach to Southend, air flip to Ostend, and thence by Blue Car through Belgium, Germany, and the Brenner Pass. Our coach driver-courier was Andre, a man of special talents, who entertained us with his singing, spiels, and travelogue data at will, as well as providing coffee at a shilling a cup made in an old oil tin, presumably scalded and clean, though the sight of such a container is offputting, somehow altering the flavour of the coffee.

We made a stop at Brussels and went on through the night. The next day found us in the Brenner Pass, with its three Dolomites peaks high up in the sky above us, when there was a technical fault in the engine and we were forced to make:

An Unscheduled Stop

So, on our weary way from Ostend, in Belgium, to Opatija, in Yugoslavia, we all poured out of the coach, famished. Andre, the Belgian courier-driver, said we would have to wait for at least an hour while he arranged to have the trouble attended to. The Ristorante was crowded and Lin, fourteen-year-old Christine, and I were lucky to squash down at a tiny table, although well away from fellow English speaking passengers. The table was really for two people, hemmed in on all sides by hungry young European travellers, voluble and all intent on grabbing a mouthful of food. It was twilight as we sat down, and the cafe lighting was soft and shed its illumination in a way preventing a distant gaze beyond the next couple of tables, the far tables offering a background of hazy vision and rhubarby chatter. *Cheddar Gorge has nothing on this pass,* I thought. Several languages were being used at once at the small tables round about us, but no one spoke English, French, or for that matter, Welsh, any of which would have been of help to us. Still the atmosphere was intriguing, and for a time we watched the antics of waitresses looking like sisters of Sophia Loren, moving haughtily round their customers taking orders and dispensing food in whatever order their fancy took them.

After half an hour, the feeding of our mind with the educational value of the experience gave way to hunger pains and we thought it must be our turn to eat. There were no menu cards conveniently available, but our mouths were watering as people all around us were knocking back food at a rate of knots.

At last, I managed to catch the eye of a waitress who brushed her way through the tight spaces between the tables and gave me a mouthful of Italian.

"Desidero mangera," I replied. My phrase-book Italian must have meant something, for she produced a menu card. I stuck my finger at two arbitrary numbers, one for food and one for *vino*. Off flounced Sophia's sister with an emission of breath like the snort of a pedigree mare let loose in a field. A quarter-hour later saw us with a dry fish concoction on our plates to clash with a carafe of red wine clearly deserving of the acquired taste of years of appreciation, together with pieces of dry bread.

Lin looked at me like I'd just failed my marriage exam. Then we were struggling to down the food as quickly as we could, but the problem of the lumps sticking in our throats was worsened by André appearing at the doorway with his right hand stuck up with the digits stretched to indicate we had five minutes more to join the bus. We were rising to leave when in came a young woman striking enough to turn all heads to the doorway. When this Cardinale type swept in, the aisles seemed to part like she was Moses parting the waters of the Red Sea. I made a gesture that our table was coming free. Sizing up the situation at a glance, she spoke, her melodic voice articulating in perfect English, "How terribly kind. I spent five years in your country, took my M.A. at Cambridge. It would have been my pleasure to help you order your meal."

I looked at the meals we had been unable to finish and the half-empty carafe and thought this was a case of sour grapes. What I said was, "It would have been delightful, signora."

At least the interlude had been something to break the monotony of the journey, and as we resumed our way through the night, the girls could think of the incompetence of the husband or father taking them on holiday. As that forlorn figure dozed off, he could console himself only with Claudia Cardinale.

The three of us really enjoyed ourselves in Opatija, where we stayed at the Grand Central Hotel overlooking the Adriatic Sea, so blue and so placid. By seven o'clock in the morning we were rowing on the water in just the scantiest of attire. In the evenings we would attend performances by Central European dancers in national costumes. In the afternoon we mosied around the area using the service buses. My school German came in handy, as I was able to talk to several of the local people. We paid a visit to the Postogna Caves, a magnificent spectacle, with the natural underground caverns accessible for a distance of three miles. A vast, cathedral-like cavern made a spectacular terminus for the light railway that conveyed visitors into the bowels of the earth. Lights went out as the train reached the terminus to switch over to return for the next trip, and in the silent blackness not a sound could be heard as parties had gathered round the guides to describe the scenes to them in their own languages, Serbo-Croatian, Italian, German, and English. Then the lights came on again, and people began to chatter and savour the sights of the stalagmite-stalactite formations and the colourless human fish alive in water held in natural basins in the walls, the products of eons of time before man.

We returned from Opatija via Trieste and travelled on through into the night to Brussels, where we stayed a night, then Ostend, Southend, and London. When we arrived at Northampton station I rang the house and asked Newton to collect us in his car, the one I had bought for his twenty-first birthday, a Rover 10.9 1936 saloon.

"Sorry, Dad. I haven't got it," Newton's replied.

"Well, bring mine," I bade my son.

"Okay, Dad."

After about twenty minutes, I rang again. "What's up?" I asked.

"Now I've backed yours into the gateway. It's stuck."

It *was* an awkward gateway. He had been backing the car down the steep slope of the drive in front of the house. We hired a taxi. It seemed that Newton's shining green Rover 10 had taken the eye of two thieving merchants who were on the prowl, setting themselves up in business, a lorry, then a car having been their first priorities, as it transpired in court after the police had recovered the vehicle, now painted black over its green cellulose.

When we lived in Northampton, we were delighted to meet again our friends Harry and Mary, whom we had met when Harry and I worked in Nigeria. Now they were back in their hometown, Wellingborough, and it was on a visit to them that we saw one of the first Soviet *Sputniks* pass overhead in the twilight sky.

It was a sad event when Lt. Col. W. B. Tyrrell, who had given valiant service in supervising contract on the motorway, failed in health with a blood disease and passed silently to his end. He had loved life so much, making no complaint in his adversity, having carried on with the final measurement of his contract while undergoing blood transfusions until confined to his bed with his terminal illness.

All staff on the M1 gave of their unstinted best, but I make no apology for singling out four men on our staff who were of great support to me, especially during the original phenomenal efforts that were asked of them from April 1958 to the day the road was opened, November 2, 1959.

Brig. Sir Robert Mariott, Kt., was my first deputy, ably monitoring the progress aspect of site works, which captured the urgency of postwar development. Here was a man who had given a lifetime of service to railways in India followed by administration work with occupying forces in Germany, later acting as a principal in an engineers' college. At seventy years of age he was still in his prime, and at the time of writing he is still going strong in his nineties.

My second deputy, William David Ross Kerr, less than half Bob Mariott's age, worked unceasingly with great application on the financial aspect, inspiring younger men in the final settlement of the thousands of tedious operations of rate fixing.

Alistair Foot, my twenty-eight-year-old administration officer, had been subeditor with the *Northampton Chronicle and Eco*, as I mentioned previously. He had the flair and enthusiasm to inspire the few administrative staff members charged with the task of committing the office business onto paper. Alistair had a great mastery of shorthand. Taking minutes of the regular Friday progress meetings, chaired five times in six by Sir Owen or O.T. and, in their absence, by myself, Alistair had the facility for getting down verbatim not only the official statements, but also the side discussions between representatives of each side, so that a more faithful record would be available as he continued, in the mould of a subeditor, offering his transcripts for my official notes done for general circulation. Alistair's training with an important provincial newspaper was such that he automatically ensured that a secretary was available to get the notes out for circulation every Saturday afternoon, with the notes on the desks of every executive along the route on the Monday morning.

In addition, the large number of official visiting parties called for a lot of personal effort by Alistair Foot. When official aerial photographic records were foiled at times because of the vagaries of the weather, it was largely due to Foot's artistic flair that we were able to turn out an amateur film record for the consulting engineers, capturing some of the action for all time, with commentary by yours truly.

Brig. F. T. Jones was a year older than Sir Robert. He was in charge of the alterations to public services, and he worked from the firm's London offices. He had the light touch of a man of experience who knew what he expected out of every meeting, giving the impression that his minutes were prepared in advance almost verbatim, such was his professionalism.

Both F.T. and I were called upon to give talks on the motorway to service authorities and engineering bodies in the north, where such people were interested to know something of the great wide way that would be coming their way in the future. In 1981, two decades later, I made a pilgrimage to the Trent Technical College, which houses the lecture room where I had given an address on November 30, 1960 to the Nottingham College of Technology, with 220 students, academicians, and many of the principals and members of the engineering departments of local authorities and private enterprise organisations.

In order to be truly resident for the extension of the M1, my family and I had moved to a house in Glenfield, Leicester. When I took my leave of that project, I had to look around for something to do. I had some interviews, including one with a firm specialising in competitive reinforced design, which was not really my scene, and a chance to go into partnership with an engineer starting a practice in East Anglia or to be a director with an established tarmac firm, none of which appealed to me at that time.

I finished up with a position as a senior resident engineer for the consulting engineers Merz and McLellan, who had a scheme starting for the reclamation of land owned by the London Brick Company Limited at Peterborough, the intricacies of which will come into focus in the next chapter, which I have titled "The Ash Pan," but as Churchill might have said, "Some ash pan!"

In the meantime, both my sons had decided to work overseas. Newton was in a bank in Nigeria, while David had emigrated to Canada. They both married when overseas. Perhaps someday they will relate their own life stories. . . .

15.   The Ash Pan

Merz and McLellan were old, established engineers. They did power stations. There were a number of these in the Midlands, traditional ones using coal fuel, and what does coal leave behind when burned? Ash. The modern furnaces produced the ash in the form of fine dust, and disposal of this at the individual stations was becoming a problem. Merz's had a pilot scheme projected to dispose of all the waste, disposing of the dust and using it to reclaim the considerable acreage of worked-out brick pits, the Oxfordshale beds having been exploited by several companies over decades of time. The whole area was now in the ownership of the London Brick Company, with its main centre of works and control being at Bedford.

When I saw the advert in the paper for a resident engineer at Peterborough, I was immediately interested. Since leaving the M1, I had tried my luck without being suited in at least three locations, but this possibility seemed promising. Peterborough was fifty-three miles from where we lived at Glenfield, Leicester, but the A47 was a pleasant country road, so daily travel was just possible. I was granted an interview and travelled up to the firm's head offices at Carliol House in Newcastle upon Tyne. Two senior engineers, S.A. Rossiter, chief engineer, and Colin Tebbutt, went through the scheme with me. At that time, Merz and McLellan had a good number of major power stations on their order books in various stages of planning and construction. In those days one thought of such a station as costing around a hundred million pounds. The Peterborough project, jokingly referred to by the two engineers as the Ash Pan, for the Midlands power stations was in fact an imaginative venture to deal with a growing environmental problem or, actually, two problems, with the power stations' waste being disposed of in an orderly manner to reclaim the land devastated by our forefathers. Neither the official name for the scheme, the Peterborough Dust Disposal Scheme, nor the domestic connotation giving this chapter its title do justice to conjure up in the mind the level of engineering thought, mechanical, electrical, hydraulic, and civil, that had gone into conceiving this pioneering work or the ingenuity and application that would be required during construction of the personnel supervising and building the first working phase in twenty months. The whole of the site activities had to take place with the existing works in production, entailing grafting one works upon another yet with a heavy penalty implicit for stopping the brickmaking.

The brickworks sites had vast areas of deep pits, some wet, some dry, with the original "callow"—the overburden—of the worked-out shale strata left in untidy heaps like the surface of the moon, all over the bottoms of the pits. The general idea was to make a water reservoir in the largest of these pits with an overall area of 476 acres. This pit was large enough to form a reserve of water, at one end, of 480 million gallons, requiring the cleaning off and setting aside of the callow, the excavating deeper into the shale to win material for building a circular dam of rolled shale six thousand feet in perimeter, half a mile in diameter, and up to fifty-four feet in height. A pumping station installed in this dam would be the prime mover for circulating piped water around the various pits, each with newly built bunds, balancing pipes, intake towers, and controlling valves, to operate the system of conveying the contents of trains of dust, which first had to be mixed with water in the coagulating channel of a new railway station built as part of the scheme with handling machinery for the purpose. The "prestflo" tanks on the wagons of these consignments from the power stations kept the dust try till its arrival at the station, their contents being loaded and discharged pneumatically. The trainloads would run around the reserve perimeter with no lost time.

The construction of the new man-made inland lake involved excavations of the order of thirty-five feet over the large area, with 7 million tons of ground having to be moved and reshaped. The earthworks in this area alone was equal to the effort necessary in the cut and fill for a twenty-mile length of motorway, and there was in addition to this similar work in forming sundry bunds to divide up other large pits into wieldy sections for controlled filling of the dust, which, because of its fineness, would take time to settle down in the water. There were sundry works like roads and rail bridges making for a challenge all round.

As Mr. Rossiter painted the picture, I became intrigued. The scheme was unique. The post paid less money than I had become accustomed to, but I took the view that it was just what the doctor ordered and agreed to start as soon as my leave from my last job meant that I would be free to do so.

On the journey up the A47 in my Morris 16/60 Rose Taupe saloon with plenty of time allowed, I stopped at the side of the road for a cup of flask coffee. It was going to be a longish run each day but, with summer ahead, not too daunting. I completed the trip, and pulled into the railway station car park, and entered the Great Northern Hotel. Colin Tebbutt was waiting in the foyer to keep our appointment, plan in hand, ready to accompany me to the site on my first day. We sipped cups of coffee, and he put me in the picture on progress to date on site.

"There is a problem over the disposal of embarrassingly large quantities of callow," said Tebbutt. "Probably the first thing you'll need to make a decision on."

When we got to the site, Michael Kennard, of Sandeman, Kennard, and Partners, the firm's consulting soil engineers, saw no objection to the soft yellow stuff lying against the new reservoir embankment, which had to be built in Oxford shale, tough, blue-black in colour, and smooth in texture.

The site staff were bungled into a small domestic-type caravan on the Norman Cross Road, the offices being provided for the contract being in the early stage of erection at a location, centrally disposed on the site, about a mile away on the other side of the main road. Deputy R.E. Ken Hooper sat on a pile of plans at one end of the tiny van. In order to talk to me he had to back out through the narrow door and stand outside. There was another door at the other end of the van for the chainman/attendant to do his attending, making the tea on "tea making facilities." Somewhere in the middle of the van, the quantity surveyor, Ian Grant, squatted, working on his quantities. Nearby was the temporary cramped office for the contractors, Lehane, Mackenzie, and Shand, with Alan Pigott the agent in charge of their construction operations. It was the beginning of things that could only get better, and in a couple of short weeks they did.

We moved into the bare shelter of the shell of newly erected site hutment buildings, squatting into more space than had been available in the van while artisan labour fitted facilities around us. On the first day there, my desk was a plank, rough-sawn, lying with one end on the floor and the other on the sill of an unglazed window, but I had seen worse when starting a job. What about the time Kearsey and I began at the Swansea Pipeline, when we shared a thorn bush? Still, big oaks grow from little acorns. There was soon established a reasonable setup for our supervising staff, engineers, soils laboratory, quantity surveying, and administrative personnel. A neat car park fronted the buildings, all in scale with the projected works getting off to a good start by the contractors, led by the purposive Alan Pigott, a man in the mould of a Montgomery, often besporting the same kind of beret in the field. Pigott's people were accommodated in another setup and compound, similar to ours, on a second patch of space made available by the London Brick Company, whose representative and liaison officer was the tall, fair, and fair-minded E. H. Burton. He would maintain an unintrusive watching brief during the course of all goings on on the owners' territory at Peterborough. A friend in need, Burton was not a man to be taken for granted where the interests of his company was concerned.

The whole site had been covered by aerial survey from which a plan had been prepared enabling earthworks quantities to be calculated from the surface elevations of a grid of points on the ground. Unfortunately, the elevations shown were not always those of solid ground, many areas being covered with water or with trees and bushes that existed in the worked-out pits. A rapid survey was required of the areas of site with declared levels misrepresented on the plans, being water levels or those of the tops of the trees. As good as were our setting-out engineers and surveyors, Allen Hunter and Alan Armitage, I put it to the firm's chief engineer, Stan Rossiter, that a crash survey operation was required before the contractors disturbed the affected areas. As a result, we drew upon W. S. Atkins's survey section to augment our staff for a time, Messrs. Shepherd and Tiltman turning up with their assistants, so that we were able to have the supplementary plans they prepared agreed by the contractors as the correct original elevations of the ground surfaces and as the genesis of measurement for the earthworks, before the major onslaught of these operations around the clock with a fleet of fourteen huge TS 24 pneumatic-tyred scrapers doing their damnedest in regular twelve-hour shifts, nonstop, night and day.

It says much for the tenacity of Alan Pigott and Peter Price, son of the earthmoving subcontractors, that the gigantic reserve was fashioned out of the previously abused Mother Earth with all the complex operations, added to unpredicted strata of mudstone rock, in five months flat, this operation being ready by the end of the earthmoving season in 1963, before the onset of the winter. The nature of the shale when compacted by the machinery was such that tests showed that the fifty-four-foot-high embankment stood without discernable settlement of its crest. On the other—east—side of the London to Edinburgh main railway line, other pits were being provided with face lifts and partitioning embankments, with 3 million tons of earthworks involved.

At a new approach road bridge site it was found necessary to dig out the remnants of a nineteenth-century bridge below the level envisaged for the new bridge in the scheme. The intended simple steel-piled cofferdam had to be doubled in depth and strengthened with diaphragm walls to withstand the enormous pressures of the wet pits on either side of the work. No record of the old bridge had come to hand in preparing the scheme. It was refreshing to be able to take that sort of problem in our stride at site level, a small change in terms of the way power-stations people relate to the much greater electrical and mechanical engineering content, civil works being minor, though here we had a big change in relation to foundation work for a medium-size road bridge.

There were lots of quite stimulating ancillary operations, like thrust boring, working on pipe bridge structures across the four track main line railway, introducing British Rail cranes, and work was always being carried out, running the gauntlet, crossing the services, underground and overhead, and the traffic entailed in the manufacture and haulage of the LBC's staple product, bricks. There was never a dull moment.

E. H. Burton kept in touch with us interlopers on his company's areas. It was vital to them that no stoppages of production occurred, but I must confess to being drawn to pop a down-to-earth question to E.H.B. when he was laying it on a bit thick, threatening to stop the contractors if they continued to behave in some matter in a manner that he regarded as quite unacceptable. It was during one of many tripartite meetings with our contractors, ourselves and the LBC, when impassive expression followed E.H.B.'s strong edict.

"Why not let us quantify the possible effect in terms of money?" I urged. "Everything can be settled with money!" I persisted, "What would be your daywork charge for the shutdown of the whole brickmaking operation?"

The other two were aghast. The threat of the unthinkable closedown had been dangled over the heads of the project personnel like the Sword of Damacles, being of such enormity that no one had quantified it in real terms. It was just something that must never happen, way above the heads of us on site. Chagrin turned to good humour again as E.H.B. became satisfied with a declaration of intent by A.C.J.P. No need for us to get personal when we were all resolved to be good boys, doing our duty to the best of our ability.

A.C.J.P. was always waiting for me when I came onto the site first thing in the morning. He usually had a question and a proposed solution to the question. When a man like him had been working through the night or from the very early hours of the morning, meeting up with an unforeseen difficulty, he usually had time to contemplate on a better way of proceeding to get round the difficulty. Cooperation between the doers, for their part and the supervisors, for ours, was first-class. It was rare indeed for work to be held up while we contemplated a modification of the contract details.

Fortnightly progress meetings in my office formed the basis of control and in holding the construction work on programme. Because of good prior preparations and site inspections by the members attending the meetings, the formal meeting was usually dealt with in about three-quarters of an hour. The numerous elements of the project always were reviewed in advance and details tabulated to show at a glance how they all stood progresswise in relation to their individual programme dates.

Points of query by the representatives of the Central Electrical Generating Board, (CEGB), Merz and McLellan, (M and M), Sandeman, Kennard, and Partners, (SKP), the LBC, service authorities and others were put succinctly, discussed, dealt with, and recorded, with all determined to cooperate. I found that chairing those meetings was one of my pleasures in life. In the end, it was satisfying that the completion date for phase one was met in October 1964, the pilot scheme and a forerunner of a major scheme to run on, with extensions, as more land would be reclaimed in the future years. In the initial years after reclaiming the land and soiling it over, the intention was to sow it with sugar beet, so from the off it would be useful agricultural acreage.

Meantime on the home front, how had we managed as a family at Groby Road, Glenfield, Leicester? Apart from a few initial weeks when I stayed in the local hotel at Whittlesey to settle myself into the job, I had found it best to travel the fifty-three miles from home to office by car, circling the city on the ring road to the north side and on through the open-country A47 road. It was peaceful to listen to the news on the car radio at seven in the morning, and Walter Gabriel became one of my idols when I was returning, always remembering to switch on "The Archers" at a quarter to seven. We lived in No.12, an end detached house and one of six new houses set back in their own service road, alongside the A50 road to Coalville and Ashby-de-la-Zouche. Our daughter, Christine, left Overstone School at the end of the summer term, 1963, to begin a course of further education at Loughborough. She was fortunate in finding new friends. One was Margaret, who lived nearby, the daughter of Basil Watkins, the owner of an engineering machinery and tool-making works in the city. Another good friend of Christine's was Veronica, daughter of the deputy county engineer, A. J. Mare. Christine had other new friends in the area and retaining contact with her colleagues at Overstone from time to time.

Petersborough Dust Disposal Project, aerial view of stage 1 pits and reservoir. *Courtesy of Central Electricity Generating Board. Photo by Aerofilms and Aeropictorial Ltd.*

It was a time when our daughter was a maturing teenager and developing quickly. She missed her brothers' company, with Newton in Nigeria and David in Canada, and so we tried to keep a watchful, tolerant eye on a child who loved to have friends, yet was the only child at home. Boys began to appear on the scene, and she began to go out on dates and help to organise parties for social events in the locality. As I said, we were trying to keep a watchful eye, with me often going to meet her when she returned in the shadowy hedges of the area in which we lived. We attended the parish church in Glenfield, though were not so active there as we had been in Kingsthorpe.

When the project at Peterborough was clearly going to meet its date for completion, I began thinking about what I was going to do after that. Fifty-three miles to work every day was not my idea of a permanent arrangement. My attention centred on a new bypass to be built at Burton-on-Trent, only half the travelling distance from Leicester and with the incentive that the near motorway standard dual-carriageway road would include substantial prestressed concrete designed bridges, a technique that had escaped my experience to date, none of the M1 structures being designed in this way by my old firm. My application resulted in an interview with Desmond Hennessey, the principal for the firm of Bolton, Henessey, and Partners, engineer for the scheme, and I was successful in being awarded the post of resident engineer, with a new challenge but with the satisfaction of knowing that I had solid experience behind me in building roads. Thanks to the sympathetic understanding of W. H. (Bill) Dixon, the partner with whom I had worked at Peterborough, my transfer to my job was to coincide with the finish of the first phase of civil engineers work there.

On the day I took my leave, it was with a convivial sendoff with a pair of treasured cuff links from Hooper and friends, a poem from Allen Hunter, and "a last cigarette from Alan Armitage." it was like leaving home, as we recalled amusing times like my having to change my 16/60 for a Cresta car because I was allergic to the smell of the finish of the upholstery in the Wolseley, wondering if my chest would always smell like that, as if something had crawled into me and died. There also had been the day when Ismay and Jack Hudson had had a go at each other and asked me to mediate, leaving my office arm in arm. Then there was the time when Duffy, the chainman, had solicited a lift in the car of a senior man from the contractor's, the latter imagining he was being courteous to the consulting executive staff, only to see friend Duffy carrying out his lowly duties in his honest Irish attire, the sort of man who mixed freely with vagabonds or kings.

Well, it was all over. Once again I was off to enter another chapter in my life, and if you care to join me so are you.

16. Relief for a Town

Burton-on-Trent was a town of character. For one thing, it was the home of Bass and Worthington—the first two words I learned to spell when I was a three-year-old living at the Ship Hotel, in Flint, North Wales. And there were names like Shobnall Fields, Horninglow, Branston, and Claymills, beautiful names that ring in the ear.

When travelling north-east on the A 38 from Lichfield to Derby, you first encountered Burton at Branston, with its concrete pipeworks. Then you had to wander on uneasily through the overcrowded network of streets, among very heavy industrial traffic, in this low-lying town nestling by the slow-running river Trent. Eventually you emerged after six miles of municipal meanderings over the hump-backed bridge over the canal at Claymills, to join the procession of heavies until they were able to sort themselves out where the road opened out on an improved stretch of the A 38, en route to Derby. The through traffic and the town were anathema to each other.

Traffic can be likened to the life blood of the country, and the Ministry of Transport had decided to provide an artery to let the blood flow freely to its destination, lest the veins of its sick member became "varicosed." They had conceived a line for the Burton-on-Trent Bypass, a new length of A 38 to be built to near-motorway standards, partly rural, partly urban in character and having to fly over parts of the town.

Bolton, Hennessey and Partners, a firm of consulting engineers, had been commissioned by the MOT to engineer the scheme, and Desmond Hennessey was the active senior partner, resident in the U.K., at the time he engaged me for the post of resident engineer for the work just commencing in October 1964. His chief engineer, Reg. Calvert, had been engaged on the planning of the work for years and was pleased to see the scheme about to be realised as work on the ground. Reg. and I were already acquainted, as he had called in to visit me in the M1 Newport Pagnell offices during the construction of that great wide way in 1958/9 and when I am sure he had picked up some points, as had the many thousands of interested visitors we had received in those salad days. It was nice to renew our acquaintance.

Travelling westwards on the A 50 was quite refreshing. It was only half as far to work as the Peterborough job, and one passed through two towns. One would not necessarily expect a town with the name

James Price, chief R.E., consulting engineer Sir Owen Williams, and minister of transport H. Watkinson reviewing progress on the M1/M45 motorway. *Courtesy of O.T. Williams.*

Coalville to be particularly uplifting, but in fact Coalville and Ashby-de-la-Zouche had equal fascination for me. One was vital and typifying the coal industrial backbone of our country, and the other had a historical background and a "mediaeviality" I associate with places like Melton Mowbray and Bury St. Edmunds.

I can now look back on the two and a half years spent on the Burton Bypass as one of my most enjoyable episodes, with the salient aspects of the scheme having to my mind, a classical quality. The Ministry concept, in providing much needed relief for the town, was unassailable. The engineers they chose were professional and enthusiastic in their approach. D. E. Hennessey discharged his commission with dignity and sincerity, introducing into his structures innovative flair in design and drawing on the architectural advice of an associate, resulting in bridges with decks employing prestressed concrete to give soft lines, curving in a manner intended to suggest natural "leaping ribbon" continuity from beginning to end in either direction. Reg., his chief engineer, with a patient period of scheme preparation behind him, was pleased to delegate all control of the contractors on site to the new R.E., as I had been accustomed to on many of my previous jobs, being fully authorised to issue variations on behalf of the engineer, though, of course, such authority, vis-à-vis the contractor, who needs to know who he has to deal with, is tempered with the responsibility for being fully accountable for such variations. D.E.H. kept himself fully briefed on essentials, with regular visits to site, holding regular progress meetings. Reg. was to

O.T. Williams; John Michie, chief project manager for main contractor; Sir Owen Williams, K.B.E.; and author. *Courtesy of O.T. Williams.*

leave the firm on obtaining a substantial post in Pakistan while the contract was in the early stages.

The contractor charged with the responsibility for constructing the works was Percy Bilton Ltd., and Percy was still in personal control of his firm, though the work in the Staffordshire area was fully delegated to Wally Barrett, the area director. What a great personality Percy was, a Tom Walls lookalike, I always said. Here was a man who had made an early enterprising start as a boy, riding a bike around the farms, delivering drums of oil for tractors, just like a Beaverbrook delivering papers. His business interests had grown as he matured, in the manner of all natural entrepreneurs until the day came when he could claim to have built a town. Not in one place but if all the work done by the firm were assembled in one area, it might well be a town. Here was the sort of man who did not forget his modest beginnings nor the men who had worked for him in his early days. Here was a man with a Rolls Royce big enough to hold fourteen people who, at sixty-five years of age was as fresh as the days he rode a bike in his teens. Here was a man who had time to chat easily not only to me as the resident engineer, but also my deputies, Hugh Webster and Roy Cotton. Here was a man who was obviously extremely happy as the head of a business empire.

Wally Barrett made the important decisions for Bilton on the Burton contract. He was a likeable chap, with a straight-from-the-shoulder technique. Insofar as Percy looked for direct results, Wally would deal with the R.E. on a job in a direct manner. When you think about it, that is the function of a director: to direct and be direct. The result was that the prices for alterations, which invariably occur on public works schemes were settled openly and quickly.

A. S. Peverall was the agent for Bilton. Pev. was a quiet man, operating in low key. He was a well-known local resident, elected as councillor for the Horninglow Ward during the time of the contract. Like Alan Pigott, of the Peterborough job, Pev. would be in to see me every day, as the work developed quickly on the site. While Peverall was the senior agent for the contractors on the site, his primary expertise and concern was for the roadworks operations, as distinct from the structures. My decisions in the matters arising under Pev.'s direct purview, earthworks, drainage, subbase, paving, fences, and ancillaries, were made in the light of information supplied by my able assistant, Roy Cotton.

Tom (T.D.) Neary of Barry assisted Wally Barrett in the contractors' Stafford office in matters structural. By the same token, R. S. Dewick was on site to organise and take charge of building the bridges and other structures, supporting Peverall with his delegated responsibilities. Tom Neary would pop into my office, often direct from Barry, his chesty smoker's cough wheezing but not detracting from his jovial good humour. Asked how he felt, the answer was always the same—"Never better." The luck of the Irish was soon to run out, as Tom Neary did not have long to live.

Hugh Webster was my deputy R.E. and right-hand man—fortyish, tall, cultured, and musical, with culinary talents. In their new estate bungalow in Yoxhall Hugh and Lois, his wife, worked in tandem to produce the lightest and airiest cheese soufflés I ever did see or taste. They had fitted in their modestly sized lounge two full-size grand pianos, with hardly a foot to spare for walking. These and other pieces of period furniture betrayed their previous Regency lifestyle in London.

Let us now plunge into my working life at Burton, peeping into the past from the time I arrived in October 1964 to the time MOT secretary Hon. Ernest Swingler, MP, accepted the invitation to open the new highway to traffic in May 1967, partaking of a few highlights still in my mind.

In most contracts, time is of the essence. In a book of this kind, surprise is of the essence, and even your author finds delight and surprise as events reoccur as they did in those days. It is, to me, like walking round Madam Tussaud's, with people offering themselves for recognition as they were known in the past.

The big old country house in Shobnall Fields provided temporary space for consultants' staff to operate from two downstairs rooms. Keith Booth, dark-haired and soft-spoken, from the firm's head office, was emerging from the house on the day I first arrived, his tripod under his arm, his chainman carrying the instrument box. Keith was the engineer-surveyor type whose expertise and intimate knowledge of the scheme at headquarters were now being drawn upon to get the setting out points established in the field, a John the Baptist for others to follow. In our highly developed society, it is no use being "near enough" where surveying is concerned. Design accuracy is dictated by limitation of land, a jealously guarded commodity; especially where buildings are involved, old or new, work has to be set out to fine limits of accuracy. When I emerged from my university life at Liverpool University, Chambers's seven-figure logarithms were the tables used for accurate calculations in surveying and other design problems. Nowadays computers permit of the feeding in of prepared programmes that may have been used before or were adapted to reduce calculation time to a fraction of that in the old days for road alignments, bridge design, or measurements.

Keith was working hard in the field, in the seven-figure-logarithm age, and on top of that, he had been acting as office manager, typist, and filing clerk. A handful of files were lying loose in the office as Reg and I walked into the house on my first day, having met up at the Shobnall Road and driven along the farmhouse path. With a plan in our possession, Reg. took me for a run round the site or, rather, strategic accessible points where something was going on. There was a swampy area Bilton was cleaning out, travelling excavating machines, and trucks on a low-level hardcore access road. Refilling with marl from the hills was to follow. Bob Robinson, the image of Victor McLaglen, our chief clerk of works was, inspecting the works as it proceeded.

When Reg had returned to London, Hugh Webster and I had a chat in the office. We needed a secretary. Keith typed out the advert. In a couple of days we had some replies. We interviewed applicants and chose a teenager, Maureen, who had experience of the terminology of the work, having worked with J. and L. Keir at the nearby Draks Power Station. She also demonstrated a very good shorthand/typing speed and commenced on the following day with the initial task of establishing the filing system as I liked it. From now on Keith could forget the correspondence and concentrate on the establishment of the centre line for the bypass with some help from Tom Carbray, a junior engineer who lived locally and who had been helping Reg. Calvert during the period of field investigations when the scheme was being prepared. Tom was knowledgeable of local geography and the historical development of the scheme, within his understanding as a younger member of the staff.

Roy Cotton had put in some years with the Road Research Laboratory and had useful experience in field research particularly in filling behind abutments on a number of roads. Roy had a good notion of what to look out for in the behaviour of soils and had a rational approach to control of the earthworks, as it developed.

Soon after my arrival, Biltons had erected the sectional huts to serve as offices for the whole period of construction of the bypass. There were offices for myself and my staff, the administrative staff, quantity surveyors, engineers, clerks of works, and attendants, accommodation being neatly arranged along the block with accesses from a continuous corridor inside the building. A separate building was erected a few yards away, across the car park to house laboratory equipment for testing the earthworks, concrete, and road construction materials. Roy saw to it that representative samples were taken of everything that went into the job, with appropriate tests and all individual and statistical criteria maintained to ensure specification requirements were being met, all results being filed for the record.

In Biltons' own offices on the same spacious square compound alongside the main Shobnall Road, Agent Peverall was esconced, with a staff of engineers, foremen, and others, according to need. It took quite a while for Wally Barrett to introduce, on site, a man to oversee the laboratory work, a responsible materials testing engineer always being called for with all the monitoring that must go on on a major contract for highways and bridges. Wally kept apologising for the non-arrival of his "chemist," a term not usual in the work we were doing, though the laboratory does perform production quality tests in the manner of an industrial chemist. The "materials engineer" on our kind of work is expected to do rather more than feed in results to production management personnel; he is usually expected to be a troubleshooter for the contractor so that work will not be condemned by the engineer's staff overseeing the contractor's operations and this abortive element of the work removed or modified at contractor's expense of money and time.

The liaison on quality control, carried out for us mostly by the Bilton chemist and staff but also by the divisional road engineer's mobile laboratory, which did tests of the asphalt layers, left nothing to be desired.

There was peace of mind, too, when it was seen that our contractor had ordered the steel reinforcement for the bridges in good time, the stack yard for the steel being in a commodious square area adjacent the carpark and fenced off with a gate lockable for security purposes. All fourteen bridges were allotted separate spaces in the steel compound.

Don Roe was a find for us—a man in his forties. We needed a quantity surveyor, and when Don turned for interview he admitted, frankly, that his previous experience had been mainly related to buildings. Thereafter, with the agreement of the London office, it was decided that Don would serve as assistant Q.S. in the early stages and we would hold back with appointing a quantity surveyor to give Don the opportunity of preparing the early monthly statements relating to the contractor's applications, always relatively simpler than when the contract works are in full spate. Don proved to be competent, systematic, and a person who worked well with the contractor's Q.S. from Stafford. Don could "pull out" figures for presentation in such a clear and concise way he would have done justice to presenting the nation's budget. After about four months, it became obvious that Don could handle the job at the office end; our young engineers, agreeing on measurements in the field and supplying figures agreed with the contractor on inclusion of Don's figures in the monthly verification of the contractor's accounts presented for payment. As time went on, with the buildup of work and related measurement, Don needed an assistant. To serve in that capacity we took on Mr. Johnson, a mature man, as technical assistant to Don, who was himself promoted to be the site quantity surveyor. Johnson proved to be of great help to his Q.S. in collecting numerous details that had to be noted in the field and collated to compute the various sections of the works in the monthly accounts.

John Philip Wain, the borough engineer of Burton-on-Trent, had been helpful in providing information on the marl hills that ran alongside the chosen route for the new bypass on the Lichfield side of the Shobnall Road intersection. The scheme included the right to excavate borrow pits at prescribed locations permitted by the local planning authority. Mr. Wain was always available for discussion of points of contact, which are inevitable between engineers responsible for a new capital works that has to be grafted on the area of a local authority's domain. Background knowledge of local people is always welcome as a starting point on which to found sound concepts of location and design, and Wain had played a helpful role during the scheme preparation stages.

I had had substantial experience that instanced the value of such help from fellow professional engineers, like that from Manchester drainage engineers, when I was an A.C.E. for the Admiralty during the early days of the war, or from the Bombay Port Trust engineers when I served as C.E. Bombay or, yet again, from the provincial engineer when I represented the Nigerian Railway Corporation at Port Harcourt and again from county and area engineers and public utilities' engineers when I was C.R.E. for the M1 consultants and, latterly, from Eric Harry Burton, the representative of the London Brick Company and himself a corporate member of the Institution of Civil Engineers and a leading member of the LBC's own society of engineers.

The engineering profession is the richer for mutual help and second opinions, freely given between brother engineers, as members of the learned institutions, although loyalty to one's employer and professional conduct preclude the handing over of data to be copied or used against the interests or without permission of one's employers. In this public service engineers tend to be more of help to engineers in the private world than the other way round.

However, local knowledge does not extend to the vagaries of the bowels of the earth, and despite boreholes taken in the marl hills having indicated an absence of bands of rock, when the borrow pits were opened up—surprise, surprise—large boulders and bands of mudstone rock caused obstruction of the scrapers and consternation to the contractor. Boulders and rocky bands had to be broken up and pulled away from the excavations. This work slowed up the whole operation of building the bypass embankments. It was possible to bury some of the rocky material and the boulders deep in the embankments below carriageway formations so that no late ill effects would ensue on the surface of the carriageways in the course of time. It called for undivided attention of Roy Cotton and inspectors at the excavation and filling sites to ensure that the integrity of the work was maintained and a fair assessment of the additional cost to the contractor could be made.

Earthworks are particularly susceptible to weather conditions, and it is usual to budget for carrying out the bulk of this work in the best of the weather April to October, inclusive, when one hopes to have a good average mount of sunshine to match the longer daylight hours, with not too much rainfall. My M1 experience of a long site with embankments built of heterogeneous mixtures of soil materials, having a variety of absorption and compaction characteristics, put me ever afterwards, on the qui vive. Contract arrangements notwithstanding. I determined to make really sure of embankment integrity. Some soils behave beautifully when dug from borrow, hauled, and compacted into embankments, which must be finished off with a nice top surface, sloping to the edges and shedding the water away. We had such a soil at Peterborough, the Oxford shale, with which we had built a solid, virtually incompressible dam to contain our reserve of water. The marl at Burton was in a sense similar, though it was pink in colour, not blue-black. Up in the hills where we dug it out it lay, well drained, with a moisture content below what we regarded as the optimum. It would stand a little wetting and still be capable of being worked without detriment to the stability of the banks so that, as at Peterborough, with continuity of operations and proper controls in relating speed and compaction, the soil went into place without let or hindrance up to the formation levels for the carriageway construction to follow as a special endeavour, with various layers of road-making materials rolled into place after the earthworks had all been completed. In the short term, embankments were protected with a prescribed thickness of extra marl, overfilled on the formation, like thatching on a hay stack. I must pay tribute to Site Clearance Ltd.'s principal, Mr. Walsh, for their conscientious efforts to carry out this work to a high standard.

It is often the case that earthworks are left to stand through the winter months. If we were certain that the rolled soil would always keep its initial strength after the carriageway was constructed upon it, all we would have to do, in theory, was scrape off the top protection layer as we went along. *But* I felt we must investigate the deterioration of the soil, if any, by the percolation of water through the protection layer. We researched the depths of softening by soaking it throughout the winter, and tests showed that we needed to thicken up on the design of the subbase over areas that might conceivably weaken unacceptably, should there be leakage through the road surface when in proper use. It was good to feel that our investigations had been fully worthwhile and done in good time.

One night I could not get to sleep, tossing and turning until half past three, when I rose to make myself a cup of tea. I was thinking about the excavation of a bridge abutment adjacent to the canal at Clay Mills, so close to the old existing bridge over the canal and tight against the busy A38 with its steep embankment rising sharply from the canal towpath. I could see a catastrophe before us. In the matter of safety, I have always put human life first. Here was a case where a strong wall of sheet piling, well strutted, was essential to be driven in to the ground before any tentative scratching about should be done in preparing the area prior to going for a base at the low level required. Any intrusion into the heavy traffic flow due to settlement of that A road could be catastrophic. The enormous vehicles passing along might well vibrate the ground enough to cause a slip. I arrived at the site at an unearthly hour, having sped down the A50 from Leicester. I was quickly joined in my vivid expression of the risks involved, with ground being cut below the main road level, by Peverall, who straightaway diverted earthmoving lorries to backfill the hole and make all safe. Maybe I was over cautious, but with all the thirty-five miles of trenches for Swanses pipeline, the bridge foundations for the M1 and the deep holes at Peterborough, it was only natural to have a sort of sixth sense at times, obviating an element of risk or was it again that guidance from my Taid?

We had to build a bridge to support the bypass where it crossed over the canal at a point in the Shobnall Fields. Two interesting occurrences spring to mind in this connection. This bridge was a skew bridge and was designed as a flat deck, supported at either side of the canal with open abutment walls allowing the passage of cattle and agricultural machinery to pass under the bypass. The outside faces of these abutments were separated from the earth embankments by a membrane of polystyrene sheets, featherweight slabs ideal for small boys to use as rafts to float down the long canal ponds. Retrieving the polystyrene was a regular thing until the earthwork backing was finished behind the bridge.

Then something happened while we were constructing the deck of this canal bridge. The idea was to place the concrete on timber formwork, supported from a space frame arrangement made up of individual cross-braced lattice girders, the frame having been purchased from Germany as available proprietary materials. Two hundred tons of wet concrete had been placed during the big, continuous operation through the night when elements developed that found the weak link in the chain. The steel frames toppled sidewise like a pack of cards, allowing the fresh concrete to fall into the canal below. Tom Neary immediately called in the Fire Service, who were able to disperse the cement using their powerful fire engine hoses, so that only clean, washed aggregate remained in the canal for simple removal by a dragline crane. All personnel on the deck had simply walked off to safety at the first creak before the wholesale collapse. On a big job, there are innumerable points of check required in order to minimise risks. The element of human fallibility is ever increasing as more and more sophisticated verification of the work is instituted, following a failure, such as that experienced in the construction of early box girder bridges, a technique dependent upon implicit observance of newly prescribed construction rules entailing site checks to a high degree of accuracy, often in arduous situations. It is like the rules of the road. At one time, if you kept to the left behind a man with a red flag, that was all you had to worry about. Nowadays there are so many occasions when the law is broken because of so many hundreds of regulations.

In the matter of safety, there is a point at which one can be so punctilious in obeying the rules that one overlooks the obvious. In civil engineering work, it is now common practice for a contractor to have a safety officer make regular visits to a site and make recommendations. This aims at making folk safety conscious. The Safety at Work Act tells everyone they must bear their fair share of responsibility. It states the obvious. There is no substitute for keeping your wits about you when you are on site, so far as your own safety is concerned. If a rogue intends to get you, you are in for the high jump. Fortunately there are few rogues, building and construction having a camaraderies second to none. Fools are a bit more common, but like on the road, it usually takes two fools to cause an accident, which brings the spotlight on me. I once did a stupid thing at Burton. Hugh and I were inspecting one of the bridges under construction for the Branston roundabout when I jumped down from the mat of deck reinforcement to the wooden working platform, about thirty inches below. My right foot was skewered by a five-inch nail protruding slyly upwards through a loose packing piece lying amongst other planks on the platform. For one who was always preaching to people not to leave nails sticking up, it hurt my pride almost as much as my rubber-booted foot. Hugh saw the problem. As I trod instinctly on the packing piece and yanked out my impaled plate of meat. I nearly fainted. He helped me as I hopped from the bottom of the scaffold across the muddy ground to our Landrover and then drove us back to the office. Brenda, my secretary, acted quickly. She prepared a hot footbath, which took care of the circulation, and squeezed my foot till out squirted the blood, in the manner of the prompt action of the school nurse when I skewered my hand with a nail through a conker in my days at Ewloe Green school four decades previously. Anyway, Brenda did a first-class job with first-aid, with the antiseptic, the dressing, and the bandage from the first-aid box. Hugh took me to the outpatients section, where I sat for what always seems to be an interminable time. There was a repeat effort by the nurse who took charge of my means of walking, with my foot left in antiseptic for an hour. The story has a happy ending, as I did not have any late ill effects, thanks to that special care; not even a corn.

One radical change that was necessary as a result of our findings at the site was more extensive deep drainage, in the gravel excavation mainly for a cutting about half a mile long at the Clay Mills end of the bypass site. Site operations and investigations revealed that there were bands of impermeable clayey soil in the ground below the carriageway level at the bottom of the cutting made in this sandy gravel area. The area had been thought to be far more free draining than it proved to be, so that we had to introduce porous side drains and a deep central drain to take the water table well below the road construction over that long length. The operation meant additional dragline machinery lined up along the bottom of the cutting, their tall jibs making them look like ships passing through the Suez Canal in convoy.

One way in which we were fortunate in the timing of the work was the amount of local hardcore becoming available just when we needed it on the site. The Midlands is not an area like South Wales, where an abundance of natural stone exists. Everyone had thought that hardcore would be in short supply. It is always a most useful material on civil engineering sites, especially on motorway-type work, invaluable for plant and vehicle access roads, pitching, drainage layers in soil, and many other uses. As it happened, Wally Barrett was able to provide broken concrete from runways being broken up to reclaim land that had been lost by agriculture to the air force during the war. Furthermore, there were a number of buildings being demolished in the town, which meant that substantial quantities of brick hardcore had to be disposed of. Favourable mutual arrangements were possible for the demolition and the construction contractors, and Wally was into this availability. The works in hand benefitted by obtaining supplies close at hand, with economy for the parties to the contract.

Let us now move along to the affairs of colleagues, friends, and family, whom I have always thought to have an important influence on the conduct of one's own life and whatsoever achievement one may realise. Certainly, in my kind of job it is only possible to see works through to a satisfactory end with their endeavours, support, and sacrifices.

Our young secretary, Maureen, was to stay with us for the first twelve months of the contract duration. Then toward the end of 1965, she was putting on weight, signifying the coming of a little one and sought refuge behind her desk with its modesty board to screen the knee hole. She began to park her blue Triumph Herald car close to the main office door and walk straight in to her seat, concealing the evidence of joy (Joy was also her own second name her full name being Maureen Joy Woods) until she went home after work.

Presently Hugh and I were faced with the task of taking on a replacement when Maureen reluctantly departed with the imminent arrival of her first baby. Husband Keith was formerly a policeman but was now taking a welding course. He was delighted at the prospect of having two new roles to play.

When Brenda Ann Dickinson turned up to work at the offices (she having been our obvious choice at the interviews), we did not recognise her as the same lady we had in mind. We need not have worried. Brenda was a girl with a flair for sporting different hair colours; each one, black, red, blonde, or what have you, suited her personality in its own way. It was to matter not what colour she had rinsed her hair, for she was always reliable and efficient at work. Two hours after her arrival, on her first day of duty, she was au fait with the system and the routine. She too was to remain a year before having to leave the job for the blessed event of bearing her second child, Mark. With husband Tom and the other member of the family, son Karl, Brenda was destined to emigrate to Adelaide.

How fortunate that when that occurred Maureen came back to carry on with the job of secretary while the baby was in the daytime care of a trusty baby-sitting friend.

A sad event was the loss of our friend Johnson, Don Roe's assistant, who was always one to read a book late in the evening before retiring to bed. It was two o'clock in the morning when Mrs. Johnson came downstairs to find him still sitting in his chair but inert, having passed away. Such an occurrence among closely knit colleagues was like a family loss, like the time Lin's Mam had passed on in the Cliff at Pembrey eighteen years before. Our London office people all shared our sense of emptiness.

When Johnson died, the gap at work was filled by the appointment of a younger man as technical assistant, David Stone. He had measurements' experience, mainly in building operations.

As the contract wore on, the bridge piers and decks were proceeding apace, with special high-tensile steel used for the sweeping long-skew span of deck to take the existing A38 road over the bypass at the Clay Mills end. That spectacular shape had called for model design to determine the stresses. The welding of bars, highly concentrated in the deck, was a difficult operation calling for careful, sound workmanship at every location, and special inspection was necessary. A gamma-ray film record was made for 100 percent of the length of the welding in order that nothing should be left to chance. That done, one could rest easy in one's bed so far as that operation was concerned.

Several of the bridges were designed to employ prestressing cables. Hugh had experience of the technique, but Des decided to introduce a new member to the staff to spend his full time on these bridge decks. He was to come over from the USA, a Dr. Thomas who was currently engaged on similar work "over there." Now when he came over, Koso, an African national, had a wife and three children, and the accommodation earmarked for them was not convenient for his good lady, who was in the medical profession at a Birmingham hospital. In just a short time the enthusiastic new recruit had settled in a new luxury six-berth caravan parked on the compound right on the doorstep of the office, so to speak, at the same time settling in to the work of monitoring the prestressing operations. After some early teething troubles, the contractor's personnel soon had the hang of what was expected, and the work went with a will.

By this time, other changes had taken place. The firm was now practising by the new name of Hennessey, Chadwick, O'heocha and Partners, as Des had now taken on new partners while remaining responsible as the engineer for the Burton job, amongst others.

For my part, Lin had tired of my returning home latish most evenings and prevailed upon me to buy a house in South Wales. So, by the summer of 1966, our home was transferred from Leicester to Llanishen, on the outskirts of Cardiff, where I had at least one famous neighbour who knew me intimately, none other than our biggest dependent, H.M. Inspector of Taxes, at Ty-Glas Road, not a cock stride away from the house we lived in. More pleasant was the thought that an old university colleague, John Hughes, was also living within a mile of us. John had worked at the Gower when I was at Swansea.

The move to Cardiff may have given solace to Lin. There was the possibility that Christine would meet up with some Welsh gent and settle down. Christine went off daily to pursue a secretarial course at Clark's College, while Lin seemed fairly happy in her new surroundings. For me, it meant going into digs and week-end travelling home. How does one fill up the time when approaching fifty and living in digs 150 miles from home? Of course, there was more time for work. Again, I was able to drop in on Hugh and Lois, Roy and Rita, or one of the others, but this does not fill up one's time. (I really should have written my autobiography then, as it would have saved me writing three-quarters of the work I am now doing as an O.A.P.)

What I did do was to take night classes and also go to the baths. What was the result? The first dive into the baths caused my left ear to block up, and try as I might, it would not clear. So off I went to see the doctor. It was fortunate I went to the right man, one who took delight in clearing out ears. After a quick peep inside with his instrument, he declared, "It's full of wax." He put some stuff in to loosen the wax and told me to "Come back next week," which I did. Then he took the tweezers and I felt a "ping" in the eardrum. The old expert produced a piece of wax the size of a pea, triumphantly, while I enquired if I would need to protect the drum with some cotton wool when I went into the open air. When he replied his voice boomed into my left ear like a town crier. "No need to bother with that," said the doc. "And it will heal up, too," which was what I was rather hoping. The wax must have developed over the period since I last had it cleared at Colombo, in 1945, twenty-one years previously. It was a pleasure to hear the metallic sounds of things that had previously been only a muffle, like changing the gear of my Morris 1100 cc, purchased after the Cresta had given me a spot of bother in Cardiff when I had been house seeking. With seventy thousand miles on the odometer I could not grumble.

My night classes were indoor activities, play reading, golf lessons, and physical training classes. The first two were in Burton town and no real strain on the system. *Hedda Gabbler* and *Murder in the Cathedral* made a nice change from drainage and bridges, and driving a float ball was easier than driving a Landrover on site. However, the P.T. course at Barton-under-Needwood shook me up a bit. As I toted my half-century-old body about in the gymnasium in competition with younger men, doing my whack on the obstacle course with its vaulting horse and again on the trampoline, my innards began to complain, and at home one week-end I writhed in agony. My doctor fixed me up with a visit to the infirmary, where a barium meal and X ray revealed my gallbladder two-thirds full of something solid. "Not worth having an 'op' at your age,"

216

I was confidentially advised, which left me a bit *suspendu au dessus de la chute*.

Lin came to the rescue when my aching body kept complaining. "Why don't you go to the health shop?" she suggested, "I heard of someone who had a prescription for dissolving away stones."

Off I went until we found such a shop and a friendly assistant quietly gave me a natural remedy. I was to starve for twenty-four hours, then proceed as follows: "Take two tablespoonfuls of edible olive oil plus one tablespoonful of pure lemon juice every fifteen minutes." This dose was to recur until I had imbibed the quart of olive oil and the pint of lemon juice purchased for the purpose.

Hitherto I had never imagined myself drinking even so much as a teaspoonful of olive oil; the thought would have been enough to turn my stomache. However, on the following week-end, what with my hunger after a fast of twenty-four hours and the fright of the pain, I put myself through the masochistic procedure over a sustained period of about four hours. Eventually the purging started, and I had the relief of seeing what looked like white sand in the pan. Don't ask me why, but that was the last of the pain. Whether it is because the stones broke up and were sluiced away or my body was just not up to complaining again, with the threat of a repeat of the same ordeal, I do not know or care, but that is what happened.

It was pretty hard going, running the job in addition to travelling home to Llanishen each week-end, returning to work at Burton via Newport, Gwent, Raglan, Monmouth, Ross on Wye, Tewkesbury, Birmingham, and Lichfield. An early start on a Monday was essential; as we were having winter fogs in the Welsh hills, driving conditions were foul. However, all went without mishap until one Friday night when I was travelling home for the week-end. Just after leaving Monmouth, the rain was bucketing down. It was a foggy, murky night, reminiscent of some of the dirty nights I used to travel from Llandeilo to Pembrey when I worked on the Swansea pipeline scheme. All of a sudden my car pulled to the off side and I was able to go with the directional change into the lighted yard of a pub on the other side of the road, pulling to rest in the heavy downpour. I had a flat front tyre and rushed into the pub to warn Lin of my holdup and predicament. It turned out that I could not get a service garage to help, and with the filthy night I was unable to try to change the wheel until it was too late. So I had to stay overnight, fortunate for the mercy of my refuge at the pub. It was just as well. When I came out in the morning it was to see that I had three flat tyres, punctures on each unlike my Port Harcourt experience, when I had three punctures on one tyre. I could not believe it until I remembered that I had run my car on the earth surface under the bypass bridge. A bulldozer had been used to backblade and tidy the earth surface and must have incorporated some nails in the topsoil. The car stood forlornly on jacks and packings while a local garage took away three wheels for wholesale treatment, i.e., one new tyre and tube, and the repairs of the punctured inner tubes on the two other tyres affected.

My half-century did come up while I was at Burton on the sixth of February 1967. Christine was up for the occasion, and we had a good old sing-song in the public house near the office in Shobnall Road. Work was coming in to a close quite quickly, and we all had the feeling of a job well done after the sustained effort during the last two and a half years, most of those present having been on the job from the very early days.

My own sendoff was to come in a few short weeks as, with the end in sight, I managed to get myself fixed up with a job as the resident engineer for a new bulk grain terminal to be constructed for the Port of London Authority at Tilbury Docks—work that will preoccupy us for most of the next chapter. Once again, my daughter was present for the big sendoff. It was like a family gathering. Dennis Calderbank, a practical engineer who had taken on the setting out since the early days after Keith Booth returned to London office, did a fine job of printing a generous token of esteem from all my staff, with A. S. Peverall and R. S. Dewick associated with the occasion. With some nostalgia, names like David Patel, an Indian engineer, and Inspectors Frank Potts, Tommy Mc Geighan, and Geoff Handford, with office ladies Margaret Killick and Gill Draper, trip off the tongue as people not given a special mention in my text above.

It had been a very satisfying period of work, in which Messrs. Emm and A. R. T. Montgomery called in from time to time to liaise for the Ministry of Transport. Monty had been a colleague of mine when we were both at the Liverpool Liver Building offices of the C.E. in Chief department of the Admiralty, with Monty, older than me, a divisional civil engineer, at that time, later to stay on until he took over the office in the Liver as the officer in charge of works. The help of the divisional road engineer for the Birmingham area of the Ministry, coupled with that of Messrs. Wain, borough engineer, Burton, and Jepson, the county engineer of Stafford, must not pass without appreciation.

Christine drove me down to the Newbury offices of my new employers, the firm of consulting engineers Shoesmith, Howe and Partners, so that I could collect my company car and be briefed on the job while she took our Morris home to Cardiff for transport for Lin and herself. Actually, I had to go up to London to collect my site car and return again to Newbury to load up with plans and office equipment for the Tilbury contract. This is shades of things to come in my next chapter,

Ernest Swingler, M.O. Transport, meeting local dignitaries when opening the new Burton-upon-Trent Bypass, 1967. *Courtesy of the Burton Mail.*

though. Before leaving the story of Burton-on-Trent, I must record that I was paid the courtesy of being invited to attend the opening of the bypass six weeks after I started the Tilbury job.

Mr. D. E. Hennessey was kind enough to extend to me the privilege of thanking the contractors at the lunch for all attending the official opening by Mr. Ernest Swingler, M. P., parliamentary secretary to the minister of transport, in June 1967. . . .

17. Fitness over Fifty

What an impact the new Tilbury bulk grain terminal was to have on me. Looking back, I see it as a change of life! For the previous sixteen years, since I had to R.E. the Clase Water Tower at Swansea, I had been concerned with works not too far from the level of the ground. Here was a scheme of new works to include a dock in the River Thames, 62 feet above the river bottom, and above that steel-framed mobile towers, each 187 feet, 6 inches high. On land the new silos to hold the grain would be overshadowed by a headworks structure 220 feet high, with the whole secured into terra firma with foundational piles forty feet in depth.

My recent contract at Burton-on-Trent had involved me in a good deal of mental agility, but this had not been matched by personal strenuous physical site effort. That sort of effort was more befitting of the average professional executive of my maturing years. The shock of trying to rectify the situation by taking up P.T. classes had nearly killed me. I was a sedentary man who sat a lot, either at a desk or in a Landrover or a car.

I had taken on the Tilbury job after an interview with G. T. Shoe-smith, a man of presence. When a project appealed to me I immediately thought of the mental challenge of creating a physical work, not the physical challenge on my person.

That, however, was to be the rub! By the time I was fifty, you could almost say that the variety in my working life was becoming monotonous. My sights had not been set upwards towards the stars, but outwards to seek new horizons. I had not been moved to seek big recognition, rather job satisfaction, and I had been satisfied by completing one job after another, and as it had happened, I had been engaged on works activities over many areas, so perhaps I had a built-in identification with length and breadth as a man who took an overall view.

When I left Newbury on May 1, 1967, I had a car full of plans and, of course, other contract documents. I was a man with a mission. Nothing had been arranged for my arrival at the site in Tilbury docks, but nothing. I was just a man in a Victor Super car OLM 320 E. I went straight to the Queen's Hotel in Grays, Essex, to put up for the night and arrange for accommodation until I could fix myself up with a second base to live at during the contract period. It was not on to move Lin and Christine from their established life in Cardiff; my mission would take me twenty

Port of London Authority; new bulk grain terminal at Tilbury Docks in early construction phase, October 1967. "Silo building" on the River Thames.

J. Price; resident engineer, fourth from left.

months, with the first ship to come in on the first of January 1969, and with the help of God, John Howard, the contractors, and others I would be through by then.

The next day I went to the dockyard and introduced myself to Ronnie Smeardon, the Port of London Authority (P.L.A.) new works engineer. My Taid was watching over me in this nautical place, as Ronnie went out of his way to ease my initial problems after arrival. First he fixed me up with a small portable office near his own considerable well-established setup. Second, he spared me the help of one of his site engineers from the offices of his assistant new works engineer, J. L. Horsburgh, whom I had met when he came to work on the new wharf extension in Port Harcourt, the primary setting out lines for the grain terminal thus being accurately established to the relief of the contractors

221

and the satisfaction of the incumbent docks engineers. This was most appreciated by myself, as I could get off to a tidy start, unloading my possessions into the office, meet my opposite number, Tony Bertlin, agent for the main contractors, John Howard and Company, Ltd., enter into my task of supervising the works from the off, and engage the services of a stenographer, Dianne, from a local agency supplying office personnel.

Piling was proceeding for the new terminal building on shore, a structure to comprise batteries of grain silos and a structural steel-framed headworks for offices and machinery control.

In a dry dock, reinforced concrete bases were being manufactured for the first group of caissons, huge cubelike structures sixty feet long by fifty-eight feet, six inches wide by sixty-two feet high. When each base was completed, walls would be built using a slipform technique, in which concrete shutters are made to slide up and up so that the concrete may fill the shutters and harden curing the time it is restrained within the supporting timbers ever creeping higher, lifted by hydraulic jacks spaced along the walls, all controlled by a console in a central position on the working platform, integral with the shuttering for the walls. They were not just outside walls. The design considerations of strength, lightness, buoyancy, and pressures called for internal dividing walls, all being constructed at the same time, a cellular construction demanding careful control. The average rate of climb was seven inches per hour, but this was increased during the warmth of the day to as much as twelve inches per hour, while at night the hardening process was retarded by the lower temperature to around four inches per hour, but the work proceeded night and day until the walls were completed.

The cribs—for that is what they were—ten altogether—had to be floated one at a time out of the dry dock, it being filled with water as soon as the four cribs in the batch were sufficiently strong to withstand the water pressures entailed. Two floats, or "camels," were attached to raise the "box" structure over the sill of the dry dock and give stability up to the point where water could be introduced, tugs being required to tow the concrete vessels out through the dock gates upriver to the river site for the new dock abreast of the silo building, there to be sunk onto a prepared gravel mattress on the bed of Old Father Thames. The new dock, comprising ten cribs in line, a total length of six hundred feet, would be completed with a concrete deck with inset rail tracks for the two mobile crane towers, erected on the site by Spencer (Melksham) Ltd. Piles had to be dollied in as a skirt around the whole finished dock to enable deepening in the future when the possibility of receiving larger bulk-grain tankers could arise. In the short term, the dock could cater to vessels having a maximum tonnage of forty-five thousand.

I have been running on a bit, remembering how I was anticipating things to come in those early days at Tilbury, with the piling proceeding on the land and the cribs being built in the dry dock. I moved quickly to find and occupy a small accommodation in the River View Park Estate, overlooking the Thames at Gravesend, so my time at the hotel in Grays was no more than a month. Each day I would make the journey to work via the Dartford tunnel, quite enjoying the routine, though it did mean a high degree of separation from my now dwindled family. There was just the three of us, Lin, Christine, and me, the girls being settled in Llanishen, Cardiff.

However, so far as my personal contacts were concerned, I had some contentment in knowing a few former friends, apart from Lin's brother Howard, who had served with some credit in the RAF and now worked in a chemist's at Clapham. There was Dave Aitken, whom I had met on the *Accra*, returning from Nigeria a decade ago, his wife and family, the home of this freeman of the Thames being at Canvey Island, and there were others whom I knew from my university and Liver Building days, more than twenty years ago. Tommy Ithell, A.R.T. Montgomery, and J. Dodd, all one-time working colleagues and now a solace to be close at hand.

When it came to the time the cribs had all been constructed and finally floated up the river, with all the inherent maritime and engineering problems of this complex operation, John Howard set out his stall for the slip forming of the silos themselves. Already they had built the heavy reinforced concrete basement that also served as a substantial foundation for the silos.

The cribs had been formed with straight intermediate walls to transmit side hydraulic pressures. The silos were circular in plan, in batteries of fifteen, with a pattern of stiffener walls, more complex than the cribs. Further, the top of the silos, when completed, would be 112 feet above the ground, reached during construction by a skeleton of tubular scaffolding and a series of ladders following up the outside walls. For the first time I began to feel a little niggle, about the sheer climbing involved. A breakthrough so far as I was concerned, a first, a personal best, in climbing to be achieved at over fifty years of age.

To begin with, there was a miss, with the walls hardening at such varying rates in different parts of the walls, so that areas of wall fell out to an irrepairable extent because the concrete had not hardened in those locations. There was the need to start again, breaking up the first few feet. Still, "if at first you don't succeed . . ."

The consultants sent down an extra engineer from Newbury office and also had Frank White fly in from Canada, where the Howe side of the partnership had had experience of building silos since the turn of the century. The original Mr. Howe was still active at over ninety, firmly in charge of his business.

Tony Bertlin was not the sort of man to give in to a setback and soon the first battery of silos was proceeding as intended, at the rate of fourteen feet in a period of twenty-four hours. I too was anxious to grow with the building and fell into the habit of popping back to the site each evening after dinner, making my number with those on the deck as it rose above the foundation, fourteen feet higher each day. The routine must have been rather wearing because, on the sixth night as I was leaving Challenge Close (an apt description), I was suddenly surprised by some clot cutting right across my bow and cursing me to high heaven. I realised I was not at my most alert, so I decided to return home and get some sleep and not burn the candle at both ends on the following evening either.

I had to steel myself to do the first climb to the top of the silos and was never really comfortable climbing above a height of a hundred feet, though, as time went on, I got better at it. It was important to have one's arm muscles developed, and I took the view that when up in the air, like a monkey, "do as the monkeys do." Confidence is essential for safety in such a situation, with a determined and deliberate effort to coordinate the eye, hand, and foot with the support. There was the occasion when I failed to get to the top and had to return to earth. This was when feathery white clouds were scurrying across a blue sky in the direction from Tilbury to the river and I was setting foot on a length of vertical ladder giving access to the top and, looking up, saw the clouds racing across the sky, with the impression that the ladder was overturning backwards above my head, enough to upset my composure, hence my decision to go to the bottom and return in a quarter-hour, taking care not to look up at the racing clouds! We can all have such moments of doubt.

I was lucky to have the help of Vic Philips as my senior inspector, on loan from P.L.A. Vic was not of large stature but highly courageous, a quality I had occasion to notice at least twice. On the first occasion, the natural division of the workers into two camps at lunch break had precipitated a near riot when one-half of the men took exception to the evil-smelling cooking of the other group. You can picture the situation yourself. The two leaders stood glaring at each other toe to toe, when work was supposed to have resumed. Vic was the senior individual on the working platform as a clash was becoming inevitable. On an impulse and not in the nature of the man or his job, it seems that Vic banged the heads of the two big adversaries together, shouting at them to get back to work. They meekly obeyed, and with other senior men soon available to take over the operations, Vic came down to earth and the office for a cup of tea to get over the shock of the obviously effective action he had taken. Timing is everything. Vic had just played things by ear, his way.

On the other occasion, the spidermen were erecting the headworks steelworks for the structural steelwork specialists Harland and Wolff. The work was easy for the spidermen. They earned their bread and butter that way; day in, day out, living in the clouds and on cold steel, driving in rivets, bolting up and torsioning up the nuts. Victor would go up and along the steelwork slowly and laboriously, secured by his harness but at a dizzy height above the ground, checking each rivet, each bolt, and each nut using a torque spanner. Finding one or two a wee bit undertightened on his first inspection had been useful in ensuring that special care was taken thereafter.

Some have a good head for heights. Jack Wright was such a one. Jack came to join me on site as my assistant resident engineer about halfway through the contract. He could climb like a monkey. He could scale the vertical ladder up to the cab of the tower crane easier than I could climb up a step ladder and took some good views of the site from on high. Jack was a young man who did invaluable work in monitoring the behaviour of the works in place.

Let us now take a break from concentrating on the work and try to relax with memories of the lighter side of my life, away from those manifold practical problems.

Lin was able to spend a few months up at Gravesend to break up the time it was necessary to maintain two homes. It seemed that our daughter was happy in South Wales, with lots of friends in the young people's associations, as well as enjoying her secretarial work. Lin and I were able to take a two-weeks vacation beginning May 15, 1968, and decided to go off to the South of France. We always preferred to make our own way and take pot luck, but it does not always turn out for the best that way. This time we were in for an adventure, the gist of which I will relate in the next few pages, a tale I will call:

The Last Bus from Marseille

At 6:00 P.M. on the fifteenth of May 1968, we arrived at gare du Nord, Paris, from London via Dover. We then filled in the time looking around the city with no set plan; our minds were attuned more to a safe midnight departure from gare de Lyons, in order to travel south to Marseille. The tickets, passports, and francs were all slotted away into the zipped side pouch of the lightweight black travelling bag I carried in my right hand.

In my left hand I carried Lin's cream coloured case, while Lin herself had our shower macs, hers white, mine blue, over her arm, together with her vanity case, the contents of which I had but the vaguest knowledge despite our twenty-nine years of marriage, which we hoped to celebrate on the twenty-seventh of the month in Marseille. We fought for time, making a detour to see the shops, the people, the Eiffel Tower, and the Arc de Triomphe and savour the whole atmosphere in one go, walking and in and out of buses, strangers in a foreign land.

Just before dark, we were sidling up to the gare de Lyons railway terminus from the south when I bethought myself of the need of a knife to peel apples on the journey, so before entering the concourse we walked round some small shops in the side streets until we found just what we required—a small, light knife with a well-tempered blade and sharp, serrated edge, but rounded at the end of the blade so firmly fixed in the yellow handle. It cost one franc, equivalent to one shilling and sixpence. We returned to the station.

Sitting as we were, with time to spare, we soon accepted the foreign atmosphere of the station. After all, strangeness is usually as to detail, not as to essentials. "Big cities are all the same," is an old saying. So are big railway stations, unless one is an expert on their technical details.

We were allowed on the train at half past eleven, and took our places in our sleeper compartment. We had not previously, nor have we since, travelled with a continental sleeper, so the unique occasion comes clearly into focus. We occupied the two lower of four bunks in the "wagon-lit" compartment and were already under the covers when the train pulled out of the station at five minutes after midnight. A portly fair-haired middle-aged man came in from the corridor and spirited himself and his bag into the bunk above me, out of sight for the rest of the night. A dark-haired young lady was already in the land of nod in the bunk above Lin.

It was noticeable to us people from Britain how smoothly the French train went as we travelled quietly into the night, with the mild, fresh, night air welcome on my upturned face. I woke to the sound of the compartment door sliding back and forth. The young girl was returning from the corridor, tiptoeing back to step lightly up to her bunk, then making her head comfortable on a pillow diagonally opposite my own, both in plan and elevation. She saw I was awake and in a quiet but audible voice commented, "Votre tête, m'sieu," at the same time moving her right hand to seal off the draft of the top window. My balding pate was indeed a bit cold; how thoughtful.

"Touché. D'accord. Merci bien," I said. Then I stuck up my thumb in gratitude and went off to sleep.

At half past five in the morning I had just been to the loo and stood puffing a cigarette in the corridor, watching the flat countryside flash by, the quaint houses becoming ever more enchanting as the gaining light allowed tinges of colour to appear. An early bird was also awake and standing beside me. She knew no English, which worried me not at all. I have always had a hankering for the French language, and the face of Michie Mitchell, our Grammar school language master, flashed before me, reminding me of the day he delivered his first twelve words to us in form IIIB:

"Six, Mille, Lille,
Alice, Achille, La Salle,
La classe, Jean, Jacques,
Toulouse, Tour, La Joupe."

It was like practising ones scales as a prelude to a piece of music. Come to think of it, Michie Mitchell, a dead ringer for Napoleon, could sing well.

The next hour or more slipped by more quickly than I can remember. The Frenchwoman was meeting a cousin and her husband in Marseille and then going on, by car, to their home in Nice. She lived in a street in Arrondissement 7 with aging parents, her mother being lame and needing a cane to walk. My new acquaintance liked going on holiday, skiing or touring or working in a relative's vineyard in Epernay a hard, physical job, ideal for a change. So said this Parisienne, who worked in a business office in Rue de Regard, under the watchful eye of a former army colonel.

I saw her shiver slightly with the passage of time in the coolish corridor, and insisted she return to the compartment. There are people we meet in life, fleetingly, never to be forgotten, and she was one of them.

My brain was functioning in English as the train pulled in to Marseille around eight o'clock in the morning. Lin and I set off from the station straight across the concourse to find the bureau de tourisme to the left on an incline across the street. Armed with an address somewhere along the Corniche, we passed the boat basin and the two legionnaires on sentry go outside the barracks of the Foreign Legion, shades of Ronald Colman's portrayal in *Beau Geste*.

The next four days, Thursday to Sunday, were idyllic. Lin loves to attend to her own food. Even our own tapwater is preferred when we go motoring around, in the interests of avoiding a dodgy tummy. In France we elected to drink mineral water from bottles. Our room boasted a large, comfortable double bed, with hard bolster and covers tucked in, continental style, a balcony to view a narrow slit of sea, and a good, solid table about three feet by two feet, with two dining and two basket-type armchairs making up the rest of the furniture. There were connected *salle-de-baines* and W.C. facilities, all very satisfactory.

Walking westward along the main street parallel to the seafront, we were able to cut through a gap in the buildings to enter onto a stretch of yellow beach to spend most of the day bathing and sunning ourselves, only returning for a bite of fresh roll and butter, and salad with coffee or tea, whenever the fancy called us. Because we were up bright and early, few of the locals or native French-looking holidaymakers were on the beach for the first hour after our arrival. Early May was rather cool for swimming, although perhaps it seemed warmer to us, coming in from a more northerly clime. Later in the day there were lots of people happily enjoying themselves.

On May 17, we took a motor boat on the very choppy waters to see the offshore island port of Chateau d'If, striking up an affinity of sheer terror with the fellow passengers tossing about on the tiny craft. Having learned about le Compte de Monte Cristo and chatted with our fellow visitors, we made a prayerful return to the shore.

We spent another day simply strolling eastwards down the corniche, in the direction of Nice. It was the year when the Rolls Royce Corniche car was launched at a price of £12,000. The whole holiday was "millionaire in the mind" stuff for us two contented codgers.

Over the week-end, the news broke that a general strike would mean no trains forthwith. What an anti-climax! "See if you can book on a coach, Jim," said Lin, and I was off forthwith, while she stayed indoors to keep out of the way of any trouble from *les personnels*, who might well be in evidence in the town. I was very lucky to get the last two seats available on the only bus not fully booked, and that would be on the following Wednesday. Putting the tickets to ride carefully in my pocket, I wasted no time in returning to the apartment. We made the most of the week-end and Monday and Tuesday mostly on the beach, not thinking it wise to be too conspicuous in all the social circumstances, but we did manage to take in an interesting visit walking up to the Notre Dame de la Garde.

Wednesday, at 2:00 P.M., saw us already waiting for the coach due out at 3:00 P.M. Looking at the passenger lists, we noticed the coincidence that another Mr. and Mrs. Price had departed on the previous day, although during our whole holiday we had met no one who was au fait with much of the English language, let alone one who claimed to be

British. Maybe the strike was a time for most of the populace to be patriotic ad nauseam.

The bus tore away at three o'clock, and we were soon heading north on the N2 autoroute for Paris, travelling on the wrong side of the road, of course. The bus ate up the road till we got to Lyons, where we left the N2 to eat in a town restaurant about 8:00 P.M. The passengers were making a bedraggled return to the coach for the appointed time of half past nine when their eyes and ears were assailed by a rumpus between a gang of *les personnels* and *le conducteur* (the driver).

Le conducteur was brandishing his paper authorising the coach to travel, protesting to the men, "L'autorité du secretaire général," while the local boys insisted, "Ça ne fait rien," "Vive le grève", and other less logical, more voluble epithets. *En fin,* at last the driver agreed to return to Marseille with his passengers and the bus, saying, "D'accord, messieurs, je vais aller retourner." So we were all allowed to board the bus, with the driver making back for the autoroute junction, not so very far away, under orders to return to Marseille, which meant swinging right down the slip road to head south. Only the artful dodger did not do that, but carried on around the roundabout and onto the carriageway heading north to Paris, with a horde of *les personnels* now receding out of sight, their shadowy fists waving under the lamplights. The driver chuckled as he turned round to his passengers, sitting behind him, one for every seat in the bus saying, "Dernier bus de Marseille," or words to that effect.

Arriving in Paris at half past five on the Thursday morning, we made our way to a bureau de tourisme to sit on the doorstep until the office opened about half past seven. With tickets from Beauvais to Lympne safely booked, we travelled home using that short air hop, with Lin consoling a teenage girl, a Miss Drew of the biscuit firm Meredith and Drew, till she left us at London Bridge station, being met by her parents, while we went rolling home . . . to Gravesend and journey's end!

It was providential that I should work in the south-east area of England for a few years. My brother-in-law, Lin's brother Howard, and I had kept in touch through all the years I had known him, since 1938. After leaving the RAF, in which he had served as a flying officer, making thirty-five sorties over Berlin and having his legs shot through with shrapnel for his troubles, he had tried his hand in an electrical supplies shop for a time before returning to a firm of chemists with whom he had worked prewar in Clapham. Howard was a quiet man, easy to get along with, with no ambition. A day's work, a quiet room, a newspaper, a boiled egg, and a cigarette and Howard was content. He did do a little private photography; otherwise I would have nothing to remind me of him, no souvenir, that is. But he had done his duty as a patriot, to be numbered by the likes of me among men like Leigh Hunt.

It was my good fortune that I was able to see Howard for a spot of lunch on a Sunday quite often during the last four years he had to live, from April 1967 to April 1971. It was a real treat for both of us to have an occasional jaunt in the car on a sunny afternoon up the Medway to watch the boats or just to a pub in Gravesend or, later in Surrey, for a quiet drink. Once we went to have a round of golf at the Mid Kent course. Howard was a natural player, having practiced as a young boy when he lived at the Cliff in Pembrey, right alongside the Ashburnham Golf Club. He had not played the game for years, but that day Howard went round the course in eighty-two, despite the casual water. I still have the ticket for the green fee, ticket no. 579 dated May 24, 1968, just after Lin and I had returned prematurely from our unusual holiday in France, the three of us having an enjoyable day out.

It was a relief that my son, Newton, was able to make good his escape from Nigeria at the time of the Biafran war, albeit with difficulty, having to return via the French Cameroons, stealing across the river Cross and flying back to Paris before coming on to stay with me for a time. It all helped to pass along the spare hours with peace of mind when I looked after my domestic needs in Gravesend.

There was no shortage of interest on the daily grind, with the grain terminal entering the last half-year before it was expected to be ready to receive the first supplies of grain. G. T. Shoesmith, the engineer for the project and principal in the partnership, paid visits to the site for progress meetings, and his senior design and liaison engineer, Don Addicott, was untiring in his attention to the innumerable points of design requiring clarification at an ever accelerating rate as the work ground on inexorably toward completion.

It was a complete surprise when, out of the blue, on July 11, 1968, I was informed of an informal visit of two very special people, a visit I recall with nostalgia and which I prefer to tell you about as well as I can in the form of a separate description of what was, for me and others, a red letter day. I call it:

The Royals' Visit

One day in the summer of 1968, when we were building the new bulk grain terminal at Tilbury Docks, with work moving along towards completion, I answered the phone. It was Ronnie Smeardon, the Port of London authority new works engineer.

"We're having a visit by royals, Jim. Can you make arrangements to receive them on your project?"

"Certainly, Ronnie. When are they coming?"

"Tomorrow," said Smeardon. "Nothing formal. They'll just be dropping in in the afternoon. If you make yourself available at the entrance to the terminal works site about three o'clock, we'll all be coming up as a party. Your end should take about twenty minutes. They'll want to get away. Oh, and by the way, tell Tony Bertlin."

Bertlin was project manager for John Howard and Company Ltd., the contractors. I rang him straightaway.

"Don't ask any questions, Tony. You're as wise as I am," I explained. "We'll stand together from five to three and await developments."

The firm that I represented as the resident engineer for the work was Shoesmith, Howe, and Partners, the consulting engineers who were responsible for the design and supervision of the construction works being carried out under contracts, civil, mechanical, and structural. G. T. Shoesmith was in practice at Newbury, Berkshire, while ninety-three-year-old Howe had founded his firm and had been building silos in Canada since the turn of the century. The new terminal had to be ready to receive tankers of grain on the first of January 1969.

I rang G.T. and told him about the visit. Did he want to come up?

"No, Jim, it's your job. You take it on. Besides, it's an unofficial visit."

Five to three the next day, the twelfth of July, saw Tony and me standing together like twins as the official party disgorged from site vehicles on the docks perimeter road about fifty yards away. As they moved slowly towards us, two VIPs were being fussed over by officials and photographers were clicking away.

"Looks like they're being presented to us instead of the other way round," said Tony.

It was true enough. We stood on a slightly elevated spot and overlooked the party somewhat below us in the foreground of the new dock extension the royals had already inspected. The rest of the dockyard premises were behind them in the distance.

I was pleased that Tony had found time to give the site a quick facelift, even though a shining green works bus with HOWARD in huge white block capitals was a bit obvious by its proximity.

"Prince Philip, may I introduce the R.E., Mr. Price?" It was Ronnie doing the honours, the London PLA officials, the police security officers, and photographers around him in a thickly packed arc. I was able to speak to the prince eye to eye, the ground being in my favour.

Philip went through a series of incisive questions. "You're the consulting engineer's man. Who actually designed the works? Are you on

Prince Charles with James Price, B.E. *Photo by Peter Coppock.*

Aerial view of Tilbury Cocks; in the foreground is the Port of London Authority bulk grain terminal. *Aerial photography by Hand Photography.*

time? When is the grain due in?" When he had received an overall picture from me, he passed on to Tony, taking up a few more points. He could see the contractor was Howard. He was aware of the London firm. Tony expressed confidence in meeting the deadline, which was comforting to me.

While the father prince was chatting to Tony then moving around, Ronnie was quick to see that the twenty-year-old son and heir to the throne was being swamped by the throng still fussing around him. Ronnie then acted to give continuity to the inspection side of the visit.

"Prince Charles, may I introduce Mr. Price?"

The young man was slim and neatly attired in a dark single-breasted suit, white collar and shirt, with the plain schoolboyish tie, currently in fashion for men. His dark, coarse, almost lacklustre hair lay down neatly on either side of a well-defined straight parting worn on the left of his head.

The prince was pleasant, relaxed, and friendly, the sort of youth you have met so often. It was by no means apparent that he was just being launched gently into public duties. He was not uptight, his modest, engaging charm becoming to his age, yet he was purposeful. I was told later that we talked for about twelve minutes. It did not seem that long, though there are few photographs in the house to confirm that it could well have been so. The photos show that we were still talking while Philip was taking in the shipping on the Thames while at the same time looking at the new offshore dock we had built as part of the scheme, with the two mobile towers a dominant feature in the river.

The big red helicopter put down near the docks perimeter road, and our visitors had to go. Charles went up into the main compartment. Philip took the controls in the cockpit, beside him the copilot who had flown in with the aircraft. The chopper rose and circled over the Thames, its occupants waving as they sped off to the palace in the manner of a flying saucer.

It had been quite an event, but it was all over.

Some weeks after that visit to Tilbury by the royals and with the grain terminal construction nearing substantial completion, I began to think about earmarking another job.

There appeared an interesting advert in the *Telegraph*. The Surrey County Council was building its first length of motorway and needed a chief resident engineer for the work. It was to be a permanent post, subject to the customary six months' satisfactory service.

In an anteroom in the county hall at Kingston-on-Thames, six short-listed applicants sat round a large polished table. It was important to me to give of my best. I had done some revision of facts and figures, but the real understanding of road building was in my head. By now, the objects of the various operations that went to make a good, sound road were plainer to me than to most, instilled by a decade in the field on substantial roads and bridges and on lots of other works having operations common to those in roads.

Each "contestant" was in for at least half an hour. When it came to me, I was face to face with three men whom I later came to know very well. There was the county engineer, his deputy, and another senior engineer. They had studied the particulars on my application, and it was soon apparent that I had the sort of experience that matched the job they had on offer. When recalled after all had been in, I was made a fair offer and was pleased to accept. It was pointed out to me that I would be expected to remain with the county for at least two to three years to ensure the completion of the length of motorway for which the start of work on the site was due in the middle of December. The reason for the gentlemen's agreement was the requirement for a local government authority to take over the commitment of a previous authority, and I had some residual pension credit regarding my service with the Swansea Corporation. I had been appointed to my first permanent post. Was I happy? I could now go ahead and purchase a car of my own again, regretfully leaving the Tilbury area, and start with the Surrey County Council Road Construction Subunit on January 1, 1969, the day the first tanker would come into the new grain terminal and the day after the completion of the contract for that construction.

One chapter had closed; another had begun.

18.   Last but Not Least

Although the river Thames episode of my life, living in Gravesend, Kent and working in Tilbury, Essex had been a happy one—exciting, stimulating, and quite satisfying from many viewpoints, the thought of moving my working base to Surrey had even more appeal. For one thing, I would be in closer touch with my wife and daughter in Cardiff, with week-end travel possible. For another, as the job to be done carried permanent status, it was possible we might get together again as a family, though in this respect, the established nature of our main home life in Cardiff had to be borne in mind and I could not definitely promise the girls more than a couple of years of stability in one place in Surrey. We would have to await events before deciding on any future move. Therefore, I would go into digs and look again for a second base to live during the initial period, selling my one-bedroom accommodation in Gravesend.

I had been lucky in my choice of a secondhand car, advertised in a London evening paper as a two-year-old low-mileage Morris 1800cc saloon. The car was a genuine find, with only eleven thousand miles on the odometer, the reason being that the owner, Mario something or other, a Covent Garden opera conductor, had spent alternating six-month periods in Canada and the U.K. Furthermore, his wife had grown a big tummy and she was finding the steering rather too heavy when manoevring and parking in traffic. Apart from needing a new neoprene bush on the steering linkage, easily done at nominal cost when I returned to Gravesend, the car went like a dream and was to give me virtually trouble-free service for the next fifty-nine thousand miles, if we exclude the broken windscreens, of which I had several when motoring on the roads in the South-east.

On the last day of 1968, I was the resident engineer for the Tilbury new bulk grain terminal, the works just completed on time. On the first day of 1969, I was sitting in a makeshift hut in Camberley as chief resident Engineer for the first M3 motorway contract in Surrey County.

Motorways in the South East counties were all directed by the Ministry of Transport Road Construction Unit at Dorking, with Maurice Milne as director. Clifford Hall, the Surrey county engineer, also had the role of chief engineer for Surrey Road Construction SubUnit, still in its infancy in offices at Sunbury Road, Kingston.

The SubUnit had designed the scheme and let the work to contractor A. E. Farr Ltd. The contractor had made a start on Monday, the sixteenth of December, 1968.

I could not have timed it better. History was repeating itself. New Year's Eve, New Year's Day, out with the old, in with the new. Remember, it had happened to me before, when I moved from Swansea to Nigeria, with the witching hour of midnight joining 1955 to 1956 as I slept like a log in a Tripoli hotel, thankful to be on terra firma. (Vide Chapter 13.)

Hall, Harvey, and Oldridge had wanted my promise to remain for a couple of years. They need not have worried. I was to see them all out, so far as service with Surrey was concerned, remaining for over eight years and would have stayed another five had not government pressure been made on local authorities to cut works. I, being at the useful break point of work well done, would be kindly put out to grass at the great age of sixty. Still, to babble on at this point would give the end of this story away, and we do not want that, do we?

Roland D. Fisher had been a rival contestant at the interview, a man of some years' standing with the county, and now he was my deputy. Roland was not a man to act "dog in the manger," and inter alia, I let him have a free reign in the matters he clearly had well in hand. A good deputy, Roland was to be a tower of support in patiently dealing with longstanding matters that had arisen during the preparation of the scheme.

The M3 had obviously been a fiddly one for the Surrey personnel to prepare. There had been numerous aspects of accommodation works to consume the time of the engineers involved with the paper work and local residents. Mercifully, the county had had the foresight to acquire and fence off the land easement for the critically important Brackendale area, built up, and through which would pass a long cutting in water-logged woodland.

When a major highway has to be contrived through a new route of country and almost always in the U.K., across a mixture of fields, woods, and urban terrain, there are a number of years of preparation to include administrative and legal procedures before even the development plan is settled. In a democracy of our sort, the most enthusiastic, pioneer-minded organisation, firm or engineer will eventually find his enthusiasm dampening. Imagine yourself playing Aunt Sally at a fairground! You've got the picture!

It is not the basic design concepts that take the time; it is the thousands of details to be considered and drawn up. Six years to prepare a scheme to nearly everyone's reluctant acceptance, two or three summers to complete it on the ground, during which time some will complain about the inconvenience and forever after it will be there for people to praise or pillory its success or failure, as the case may be. The general concepts of the M3 scheme were sound. However, when work started on the site, there was quite an amount of detail still to be gone into, and much had to be done by staff on site, with calculations and detailing of complex alignments having to be worked out and drawn up. This put staff under pressure, because the very men with responsibility for supervising the work also had to be finalising design work—a dual role. Inevitably it would be necessary to issue a large number of instructions for the details to be incorporated in the works.

All extensive site areas are prone to surprise. With all the will in the world and all the site investigation permitted by even the most cautious developer, ground is still so perfidious. The M3 problems were known to be tough, and the contract had to cater to long lengths of two-stage dewatering operations in order to draw down groundwater prior to excavating deep cuttings and laying permanent porous drains sufficiently deep to establish a new low level for the groundwater table, well below the carriageways. In chapter 16, I referred to something similar being done at the Burton-on-Trent Bypass.

The excavations at Brackendale entailed careful selection of the material arising, using it for embankments or throwing it away, at the same time evaluating in practical lengths the strength of the new surface of earth prepared to support the carriageway construction, with its prescribed sheet membrane, its granular subbase, the base, and the asphalt layers. As is so often the case with earthworks, despite the boreholes that had been drilled in the ground to give a guide on what to expect down below, we were finding what we should not have found and not finding what we should have found, resulting in giving continuous control and directions to the contractors, with investigations in advance of the particular operation so that work should not be delayed unduly. The tempo of this surveillance is grist to the mill of the professional site engineering staff and when the work is moving apace, a great spirit wells up on the site, and the supervising engineer's staff seek to get just what is required and the contractor's staff seek recognition for the work they are called upon to execute. Records have to be taken by tape, level, theodolite, camera, and field book to enable documentation in support of agreements made or if not, presentation of a claim for payment.

As is customary, staff and accommodation had to be deployed to dovetail with the contractor's organisation, initially under the leadership of Ian McLennan as project manager. The contract length in the western area of Surrey lent itself to a natural division into two sections of work, split geographically at a point to the south of Brackendale cutting, mentioned above. The two sections were organised as complete in themselves, with agents Ray Weaver and Graham Meek responsible for the conduct of the earthworks, the dual three-lane carriageways, the hard shoulders, the drainage, and miscellaneous items, as well as the structures under and over and the moving or attendance and liaison on the affected services belonging to the statutory undertakers.

My trusty assistants Roger Pullinger and Alan Budd, the section resident engineers, deployed their own assistants in turn, to dovetail with opposite members of the contractors' organisation.

For general testing we had a central laboratory in our C.R.E. offices' compound, the lab being under the wing of another senior engineer, Dick Burfitt, who later gained his Ph.D. and also acted as my section engineer for the asphalt operations for the whole contract length. Roger had a good head for structures and was assisted by his A.R.E., Alan Hutty, while Alan Budd was much occupied in the early months completing the accurate details for approach roads and interchanges, with the sterling support of assistant engineers John Marshall and Diepen Ghosh.

As month followed month, despite all problems, the two-and-a-half-year contract proceeded with an eye on progress related to programme, with a finishing date of June 16, 1971. Farr was a family firm with a reputation for good work, and in this regard R. G. Weaver was certainly one to follow in Farr's footsteps. Paying every attention to the finish of the structures in his section. He was one to cooperate with our Harold Evans, a clerk of works who was exceptionally conscientious in his job, giving useful help to the contractors' people, by reason of his own experience in the field of quality shuttering work. I give special mention to Harold. One day, at home in Farnborough, he suffered a heart attack and passed away suddenly, which was a shock to all on the site. Remember Johnson? For me, it was Burton over again as Evans was one of the people who gave of his best. "He had to do with roads," said the vicar. *He has crossed over his last bridge,* I thought of his work on site. "A proud man," said his colleague Bill Duncan, speaking to me after the funeral service. "Always on the go." "A much loved man," I hereby recall for posterity.

During the first months of the construction of a major highway site clearance, fencing, and removal of silt and unwanted materials progress along with the early excavations, drainage, piling, and bridge foundations in accord with the resources on the site predetermined to meet the requirements of all concerned—a good job, to be completed economically in the contract time. At M3 in Surrey, dewatering was a major design feature of the temporary works and it took a long time to get through the earthworks. In the first winter, we were all virtually in the mire from end to end, with little or no means of access to most points of the wooded, marshy site.

Some of the structures were of prestressed concrete, and up at the Lightwater end roundabout nonstandard bridges had been designed by a lone member of the Surrey SubUnit staff when at Sunbury and later at Guildford, a gentleman named Diepen Ghosh. It was deemed expedient to have Diepen work at the site, simply to ensure that the particular bridges were properly put together, a wise precaution, as rechecking the design was done on site, before the Lightwater bridges began. This was before what is now a compulsory requirement. Nowadays bridge designs have to be checked by qualified independent designers.

Problems of earthworks included the need to remove an unexpected quantity of unusable soil and, inter alia, to bring in significant quantities of gravel, the deposits in the ground having proved to be unpredictably low. The preponderance of weak soil necessitated a granular layer on top of the embankments sufficiently thick to ensure a uniformly firm surface on which to build carriageways that could withstand the loads of the heaviest vehicular traffic without unacceptable deformation.

In cuttings, the variability of the ground strength meant that cross-drainage and surface stabilising of plastic soils had to be done in individual local lengths, laboratory technicians being always busily engaged taking samples on the site and testing them in the laboratory.

Throughout the progress of construction for a work of this nature, there must be careful monitoring of the prescribed standards for every engineering operation, whether in drainage, in earthworks, in the carriageways, or in the individual stages in bridge building.

Stressing is a technique used for the design of some reinforced concrete bridge decks or in other parts of a structure where potential weaknesses would exist using just plain or normal reinforced items. The philosophy is to use cables to help concrete's weakness. Its ability to withstand tension unaided is only a tenth of its ability to withstand compression, so by helping it by moving the datum point of stress below zero, viz., inducing a prestress, more versatility is available in design. The operation of prestressing on a site must be monitored with special care having regard to the enormous forces to be set up in cables by hydraulic jacks. A broken cable can cut a man in half should he be standing in line with its flight. A careful programme of checks is essential to ensure precision to attain stress objectives, which must be locked into the cables by anchoring and grouting up solid into the concrete, the whole process being a sophisticated professional and practical challenge.

The running surface of a highway is vitally important. In our country over 80 percent of surfaces are blacktop, i.e., asphalt, bitumen macadam, or tar macadam, supported underneath by layers of well-rolled or otherwise compacted stone or gravel, often in matrices of cement or bitumenlike substances, all intended to present a solid support and true for the service expected by society. About one-sixth of our new running surfaces are provided by reinforced concrete, laid to a fine degree of accuracy. Thicknesses and specifications of the various layers are decided in accordance with proved design criteria based on the traffic to be sustained by the running surface. The M3 contract, which ran from the Surrey/Hampshire border at the south-west end to the Windlesham Brook beyond Lightwater at the other, was to have an asphalt running surface. Someone had to take on its monitoring. Who?

The section resident engineers. Roger and Alan, each felt fully committed with their work on hand, and in view of the testing content in supervising each layer of rolled material, it fell to Dick Burfitt, the materials engineer, to take on this section of the contract with the conscientious assistance of Mike Walton and the never flagging laboratory and field technicians, Steve Harris and Terry Howells and others. It is to the credit of all concerned that very few and very small areas required to be rectified by reason of being outside the parameters of the specification for material or workmanship.

Besides the engineering staff required to cover the different disciplines of the work, the amount of quantity surveying was considerable, for the size of the contract, which with its long approaches at three major intersections and the numerous changes dictated continuous application to the measuring up on the site, the taking off from drawings and evaluating the works as done for certification of appropriate sums due to the contractor at the end of each month. W. J. (Vic) Bridges and Ian Small, his assistant, performed with credit in this work.

Over the period of the contract the story unfolded on paper in great detail as ten thousand letters, including two thousand formal instructions, as well as masses of minutes of regular weekly site meetings and ad hoc meetings, entailing a busy job for the administrative personnel.

The more staff on site, the more detail is committed to paper and record, and with practically the same number of staff in the field as I had had to assist me when C.R.E. for fifty-five miles of the M1, done in 60 percent of the M3 contract time, the paperwork was comparable. The quantity of physical work in the field was but a fraction of that Great Wide Way, but the level of the overall achievement was still considerable.

In order to monitor the general progress and conduct of the scheme, the county engineer in his role as chief engineer for the subunit, would hold a meeting at the site at intervals of approximately one month, ensuring human personal contact between the parties to the contract, so far as one can expect in this age of anonymity. At such a meeting, as C.R.E. one can report to an individual, rather than an institution, and the manner of one's report is tempered by the presence of opposite numbers who are the contractor's people, equally concerned to see the job proceeding satisfactorily. While the brunt of accountability by the contractors must be borne by the senior representative for the contractor, it may well be that specialised parts of the work like piling, asphalting, prestressing, and even earthworks or individual bridges, if sublet by the main contractor, call for advisory people to support the contractors' front man in giving a comprehensive report of their side of the picture of the contract, its progress and conduct. For the vast majority of the time, David Wallace represented the contractors at the site, with the title of project manager, which was just as well, for the transient nature of our society was destined to change the Ministry of Transport into the Department of the Environment, A.E. Farr Ltd. into Bovis Construction Ltd., and subcontractor Dick Hampton some time later was to succumb under the umbrella of Cementation Ltd.

Bill Pomphrey was an early one to sign the visitors' book. Bill was a project engineer at Dorking, where existed the offices of the director of the South East Road Construction Unit. When Bill arrived at our newly erected sectional building offices by the A325 at Ravenswood, Camberley, on February 28, 1969, he did as he would always do on his visits to sites in the area for which he liaised for the director, Maurice Milne. He looked in at reception and had a word with Kathy on the switchboard, at first our girl Friday, later acting as officer manager when we took on secretarial help. Kathy would announce his arrival, and after a few minutes' chat with me Bill would get down to the nitty gritty of longstanding paper problems with Roland, my deputy, who was equally acquainted with the landowners and local drainage and services and who was still in the throes of wrestling with perennial queries arising.

Visitors found that the foyer to the offices had the pleasant touch of a female presence. Of course, reception work has to be related to the size of the staff and number of visitors. We did not have to receive anything like the number of visitors on the first Surrey motorway as on the first national motorway, M1. Between 1958 and 1959, such work was in the pioneering stage, with the rate of construction phenomenal. Here in Surrey, a decade later, with hundreds of miles of motorway already now under construction and completed, the M3 was merely regarded by most locals with bland tolerance, mild compliment, or in some instances adverse criticism. However, we did have the pleasure of quite a number of visiting parties, mostly interested engineers, councillors,

Round Tablers, and various. They were people with general or specific interest in the work, including business objectives. When such folk turn up in reasonable numbers, they present a pleasant diversion for whoever on site was best suitable and available to escort them round the parts of the work of particular interest to them, which is quite often the lot. We had people from as far apart as India and Argentina.

As to my personal arrangements, by May in 1969 I had fixed myself up with a new flat in Hill View Road, near the town centre, in Woking, a convenient second base to live in, in a civilised manner, while doing justice to my job. Occasionally it was possible for Lin, Christine, or Howard to pay a visit. More frequently I would return to South Wales for the week-end and return to the job on the Monday morning. I found it expedient to start out at four-thirty and arrive about a quarter to eight, rather than leave half an hour later, meeting the traffic towards the end of the trip with a much longer travel time. So the time passed with no thought of the family moving up to Surrey for the best part of two years from the time I arrived on the Surrey scene. But something was to change all that way of living.

During the Christmas holiday of 1970 we were all enjoying ourselves as a family: Lin, Christine, and Howard. It was a jolly time when both our near-Cockney residents were in our native Wales, taking the girls for a good run round the countryside of South Wales.

"I don't feel much like going back to work," Howard said when Boxing Day was over. This was unusual for him, a man with an innate sense of duty.

"Neither do I, Howard," said I. "We've been spoiled down here."

Howard returned to work on Tuesday, December 29, intent on doing his stint at the chemist's shop in Clapham. I returned to Surrey on Monday, January 4, 1971, inviting Howard up for lunch in Woking on Sunday, the tenth. We had a trip round in the car before lunch, as we had done so often previously, calling for a drink at a roadside pub near Chichester. Back in the flat, lunch was my usual affair, grilled steak, and two or three vegetables, boiled in one pot with a blob of butter to make them palatable. As we ate our food, Howard was spilling some of his at the right hand side of his plate.

"I must get these glasses checked," he said. "I've had these new ones, but they don't seem right to me."

About half past four, I saw Howard off on the London train, hoping to see him two weeks later, as I would be in Penarth on the following week-end. However, on Monday, the eleventh, Howard rang me from London. "Thank you for a great day out yesterday," he began.

"I enjoyed it too, Howard. How is everything?"

"Not too bad. Having a bit of trouble with my eyes," he said. "Got to go into hospital for a checkup."

"I'll come up to see you right away, Howard."

"No point, Jim. Nothing you can do. Just having a checkup. All the best. I'll ring you later." He put the phone down.

I felt a little anxious, but apart from an intention to ring up Mr. Dinsdale, Howard's employer, in a day or two, I did not attach tremendous importance to Howard's comments. I got down to work.

A couple of days later, I received a call at my M3 site office in Camberley. It was from Howard's employer.

"Is that Mr. Price? Are you Mr. Williams's brother-in-law?" I confirmed that I was, and he went on, "I'm Dinsdale, Mr. Williams is still in hospital. I am afraid the news is not good. It is serious. He has extensive cancer."

He gave me the name of the hospital and the ward. I was distraught. In life there are times when our road is smooth, others when our way is a bit uneven. There are highs and there are lows. This was an abyss. I immediately thought of the Sunday lunch and the fact that he had mentioned to me his recently taking a tablet to ease a stomach pain. Being a chemist living alone, smoking forty cigarettes a day, he had wondered if he might have had cancer, but the pain had not recurred for some weeks or, if it did, he had said nothing about it.

I rang Lin. She said she would come up immediately and stay at the flat at Woking so we could pay visits to the hospital. Back at the flat that night, I mentioned the news to a close neighbour who lived in the same block.

"He seems like a chap who could stand a lot of pain," said Mr. Shepherd, my neighbour, and our mutual friend.

"Indeed he is," I agreed, thinking of the time my brother-in-law had had to drag himself along with a leg full of shrapnel, after being shot down over the Channel while he was returning from one of his many sorties over Berlin.

That night I went on the train to see Howard. The journey is like a still-life portrait in my mind. In the train a young woman sympathised. In the reception hall of the great St. Thomas's hospital, a janitor and a nurse were gently dissuading a tramp from forcing his unwanted, disheveled presence into the building.

When I saw Howard, I had a shock. He had been receiving radiotherapy to his head. His hair had almost disappeared and what there was lifeless wisps of white. A youngish lady sat at his bedside.

She introduced herself, "I am Mr. Williams's colleague at the shop," and gave me her name, which probably did not register at the time, for I cannot recall it.

"How are you, Howard?"

"Fine now, Jim, much better," he said. He could not tell me much, nor did I expect him to say anything. When I had last seen him he had the look of one fifty-nine years old. Now he looked ninety-five.

A specialist came in and, before leaving asked me to have a word with him later. Presently the young lady made to go. She looked down at Howard with such sympathetic sadness that I felt more for her than for my own depression, filling me with anguish. She kissed him lightly on the cheek, then left the ward.

During the next few weeks, Howard received more treatment, with Lin and me paying regular evening visits until the news came that he was to be discharged. They had done all they could there. It was a terminal illness, and he would be happier at home.

Lin and I now saw to his uncomplicated affairs. He had few worldly goods, his flat in Sisters' Avenue a humble rented room, the bottom of the single wardrobe stacked with newspapers, folded and collected like others might collect journals. He also had a small ring to boil an egg, a single bed, and a leather valise. A lightweight mac contrasted with the neat dark suit and dark jacket with neat clerical grey trousers, with a couple of shirts and ties to complete his wardrobe. A camera with a few self-portraits spoke of his hobby to counter his loneliness. A wallet with a few pounds and a photo of a young lady friend long since married had already been handed over to us by Mr. Dinsdale.

Lin is thorough when it comes to cleaning. The room was cleared out and scrubbed before we handed over the key to the landlady in a downstairs room. An old gent coming in from the street met us as we went out. He told us of Howard's kindness, how he had chaperoned the old man home when they had bumped into each other some miles away from Clapham to ensure his neighbour was safe. We loaded up the few things in the Morris and returned to Penarth, where Howard would have as much peace of mind as we could make available. A few days later I drove my brother-in-law to South Wales, my old felt hat on his bare head to keep him comfortable. It was nice in the car, and he enjoyed the sunshine of his last car journey.

As the next few weeks wore on, our patient became progressively weaker. There was nothing the family doctor could do except keep Howard free from pain. The morphine tablets were there, but Howard would hold back as long as he could, then break one in two to lessen the dose when the pain became too much to bear. Shepherd was right. Howard never complained. One of us had to conduct him to the toilet, as he moved, foot by foot. I only saw the agony of the situation when I was home at the week-ends.

"You would not think a man could get so weak," Howard said to me. Another time he confided, "What a way to go."

No one ever spoke of Howard's complaint. The brunt of the work fell on Lin's shoulders: cleaning, shopping, the shops being some distance away from the house. The brunt of the nursing fell on the visiting nurse, who was strong and able to make the bed and make Howard comfortable. The brunt of the task of having him attend to the calls of nature fell on Christine, now twenty-three and still living at home.

On April 2, Howard called out. He had been sinking quickly and, with his last waking breath, said, "Hold me," while Lin, Christine, and I squeezed his hand in our three all at once while he fell asleep. During the night, he passed away. On April 8, he was buried with his mother and father in Pembrey Churchyard—a son of Pembrey. I have known no greater friend.

A few days later, in the line of duty, I was sent by the South East Road Construction Unit to attend the 1971 motorway congress "Today and Tomorrow" held in London at the Royal Lancaster Hotel on April 26–28. Some seven hundred delegates attended, which was just about one delegate for every mile of motorway built in this country at that time or about fifty miles per annum since construction first began in 1957. This was the same rate of construction as had obtained from the first pioneering contracts, predominantly connected with the M1. The governing factor for building has seemed to be the limitation of the speed at which administrative procedures move through the bottlenecks of our beloved bureaucracy, certainly not the capability of our engineers or contractors. At the same rate, the two hundred thousand miles of public highway in our country would have taken our forefathers four thousand years to build, which only goes to show how much better they were than we are.

Of course, the congress was most useful as a recap of ideas and progress up to the minute and as envisaged for the future. Sage saws emanated from men on the rostrum, university professors, ministry officials, consulting engineers, chief constables, workers for fire services, members of statutory bodies, and spokesmen for the many interested people whose livelihood moved them to participate. The big meeting moved inexorably forward on programme, as speaker after speaker introduced papers prepared by their underlings—all very formal. Light relief came when one gent spent his time with a cartoon exposition of "minimotorways," to cause a titter or two to escape the mouths of engineers with a sense of humour. No mention was made of the late Sir Owen Williams. This struck me as odd, as he was second to none as an engineer pioneering our national motorway programme in the immediate postwar period. His endeavours were manifest in works in many counties in England, Wales, and, earlier in his life, Scotland.

More memorable to me than the congress itself was the news I gleaned from former M1 colleagues who attended that day, most of them looking greyer and more serious than when I had known them eight years before. Alistair Foot was dead.

Alistair Foot, office administrator at Newport Pagnell on my C.R.E. staff, had been a great man with words. Joining the team from the *Northampton Chronicle and Echo,* where he had been subeditor, Alistair had fitted the bill of getting out the paperwork on time. At meetings he used to take verbatim notes, his speed in shorthand enabling him to get down even the asides by the meeting members. He was instrumental in publishing the minutes like a newspaper every Saturday by about 4:00 P.M. so that staff along the motorway section offices had the weekly edition spelling out the engineer's decisions at first hand. Again, in the dreary winter months, when aerial photographic records were difficult to obtain, he had helped me put together a movie representation of the works designed by the firm and in progress. He had scripting fervour. In December 1963, I had a letter telling me of his hopes for his writing career, which was beginning to take off. But as I was to learn much later in my life, when Lin and I went to the Strand Theatre in 1981, Alistair had passed away in 1971 during rehearsals for the play for which he was coscriptwriter with Anthony Marriott, the longest running stage comedy ever in the world: *No Sex Please, We're British.* He could not have been more than forty—a premature death—so I left the congress with a feeling of sadness to tinge the interesting review of the motorway era.

While Howard lived, keeping a house in South Wales seemed to have a special point for Lin. This I could understand. It must have been an implant into her mind from an early age, when her father had taken in her mother's brother, Roger, a bachelor, to live with them at the Cliff, where he had lived until he had died. Lin and I had come to terms with the thought of retiring in a few years' time, with Howard having a home to share, an expressed wish of Lin's father, had Lin remained single and at the Cliff, in Pembrey. Such wishes are binding in honour, and the technicality of our being married would not abrogate the wish of David Williams that his son should not be homeless. When Howard died, the obligation no longer applied. Lin's first reaction was to get away from the house in Penarth where she had suffered bereavement, so we straightaway put it up for sale in order to settle in Surrey. With Christine needing to work, the pressure was on to find somewhere in the town, with Guildford preferred, to replace our home in Penarth. It would sever connections, but after the trauma, neither Lin nor Christine thought that so important. The house in Wales was sold and another earmarked in Onslow Village, Guildford, but in the interim we were all together in the flat at Woking, all surplus furniture going into store or in the garage. This happened at the end of May, by which time the contract for the M3 was nearing its due date for completion—mid-June. Christine managed to get herself fixed up with a secretarial job in West Byfleet, soon finding new friends, though she still had a special boyfriend in South Wales and spent much time on the phone talking to him, at first, that is Later the friendship waned as new interests took over. At the back of her mind she had a compulsive desire to take up nursing, having witnessed for herself the paramount importance of that profession, but for a time she would have to be patient before being admitted as a student.

Back at work, the finalising of the outstanding items was in sight. Staff had prepared long lists of such sundry works essential before the works could be declared "substantially completed." Peter Braithwaite was now the contractor's representative at the sharp end. He kept his people nibbling away, disposing of the items on the lists, as he walked round the site, seeming to be talking to himself. It was not that. He was talking into a portable tape machine to make sure that the things spotted by his eagle eye were noted and typed out in the office.

Bill Pomphrey still kept popping in at the office, but now with more frequency as the project engineer from the Dorking office was concerned to know exactly when the road could be opened to traffic. The date for traffic to flow would apply to the Hampshire and the contiguous Surrey section. Furthermore, Bill kept pointing out, the date had to be published two weeks ahead, as it was a statutory requirement. Once the date was published, the new highway had to open! Whew!

The works began to look shipshape, and by the beginning of June all looked set for opening by the seventeenth—the contract date for takeover from the contractors. Bill fixed the date as near as possible to suit administrative requirements—June 18, a Friday. Footling little jobs still had to be done along the route and grass had not established on all the slopes, but all seemed feasible. Then with just a few days to the deadline date, came the deluge: heavy tropical rainfall. There was no time for niceties of inspections by several people, with ensuing reports and typing and orders and assembling of gangs in a nice, prearranged manner.

"Peter?"

"Yes?"

"How about coming with me round the site?"

"Okay. Certainly."

Off we went and in very little time Peter and I had covered the whole site, he noting his problems and, at the same time, being possessed of direct instructions for any shortcomings of design to cope with the catastrophic precipitation on a site where the indigenous soil was silty sand, vulnerable to water damage in the first flush of seeding. Washouts called for an amount of extra cutoff drains, to protect the site, and cleaning of sand from the shoulders was inevitable. To the credit of the contractors and our own inspectors who worked hard to verify the cleanliness of the chambers by lifting the heavy covers, there was a happy ending. There had to be! What about the arrangements for the opening? This was to be simultaneous at the Hampshire and Surrey ends. In Surrey, the traffic would be let down the slip road from the A322 between Chobham and Bagshot. In Hampshire, a similar arrangement would apply at the south end, where traffic would be let on to travel north from the A33, where the new motorway section would begin. There was to be a buffet at the south end, but as far as the Surrey length was concerned, the opening was to be entirely without formality.

The C.R.E.—me—had to hand over the new highway to the police—Vic Drummond, chief inspector, who resided opposite me in Hillview Road, at Woking. Typical of the men in the force who assist in the smooth running of the traffic throughout the period of construction of such a work, a task so often for granted by the general public, Vic had been to site traffic meetings many times. We were well acquainted.

What about the day? It was a grand day, but Vic and I in the Panda car were not too concerned about the weather, more about whether any bricks or other impediments lay on the carriageways. We made a couple of sorties starting quite early in the morning. A few bricks lay on the blacktop, heaven knows why. An assistant engineer on loan from the Ministry of Transport was tinkering with a screwdriver, checking connections for electrical switches at the side of the shoulder. It was Bob Jenkins, a young man with an alert, inquiring mind. Bob always had a way with things and people, a patient driving instructor of army territorials during his special leave for summer camps and, not surprisingly, to demonstrate when he was a tennis umpire in later years, how to reduce to low key the histrionics of John McEnroe by concentrating on the man's professional tennis and keeping him on the parameters of acceptability. "No problems here," was Bob's report, and off Vic and I went to see what else was afoot.

Down at the Frimley end, the Blackwater Interchange had a problem. The signs were not complete. Dammit. It was eight-thirty. You cannot do without signs.

"Not to worry. They'll be ready by eleven," said the man.

Neither Vic nor I were nailbiters, but we did a bit of fast moving during the next couple of hours. Eventually all seemed serene with only a few minutes to go to eleven o'clock. Down the slip road, on the wrong side, at Lightwater came a car with Bovis the contractor's people arriving from their H/Q. Even the builders are in the same bracket as the general public when a motorway is opening, so they had to retrace their steps to watch the opening with everyone else, standing on the south side of the Lightwater roundabout, looking southwards down the M3.

"The road is yours, Vic," I was able to declare at about five minutes to eleven. Now it was necessary for police radio communication between both ends to verify all was okay for the opening.

As I have said, there was to be no official opening, but what was this? At two minutes to eleven an elegant Rolls Royce swung on to the overbridge from the direction of Chobham. The chauffeur drove sedately. In the back was the regal figure of a young princess wearing one of those distinctive hats she used to wear in those days, with its large brim up in the back, down in the front, and golden yellow in colour. Whether she knew it or not, Princess Ann had graced our opening on the way to Ascot Races, although, as decided, there had been no formality, but all was well.

When a motorway opens, it is in new hands. Those of us building the motorway had been, so to speak, on the inside looking out. Now we were on the outside looking in. Any future inspections and measurements to verify the completion of sundry works and measure up for the final account would be by the courtesy of the police and the highway authority whose personnel were delegated to attend to the maintenance, in our case the county's area engineer office in Bagshot.

Some teething troubles occurred. These were related to drainage. An outlet chamber intended to work sedately, decanting clean water, separating it from the oily liquid in the effluent of the carriageway runoff, just blocked up with sand after heavy rain. It was a folly that had to be modified.

A year of maintenance was not something likely to satisfy me, and I was offered a small dualling job to do at the same time, while Roger Pullinger took on the sundry works in dotting the i's and crossing the t's, so to speak, for residual side road and accommodation-type items, not such as to prevent the use of the motorway, together with the minimal remedial work involved. The Bulldog scheme at Ashford Middlesex was also to occupy me for the next year or so, dualling a length of the A30, while still keeping an eye on the M3 during the twelve-month "period of maintenance" before finally dealing with the claims that would ensue to bring to account the special circumstances of the detailed instructions "during the currency of the construction period."

I was now back with more emphasis on the physical, the practical, and less on the engineering, the sophisticated side of highway engineering. For a year I was to be the whole engineering complement of the R.E.'s staff, with Harry Fryer as my sole help, my clerk of works and "head cook and bottle washer" so to speak, as we shared a small caravan office. Actually, it proved to be a most pleasant time for me, a man of fifty-four, as I had to get down once again to setting out works, drawing, writing all instructions, and attending to all office matters. It called for an easy, rational approach. From ten thousand signatures in the last two and a half years during my supervision of the M3 I was now to find it necessary to issue only about one hundred written direct instructions, to control the contract. Norman Hughes, agent for contractors John Mowlem, was the cooperative recipient of those orders. Norman had spent years doing about twenty similar jobs, with each one noted in a neat list in a special notebook. He was a specialist in his field.

Harry Fryer was as good with kerbs as clerk of works Bill Dalton and chief inspector Ernie Jones had each been with pipes in the good old Swansea days, related in chapter 12. Harry had worked with "the Middlesex." Middlesex County had been absorbed by Surrey and others as a result of the work of the Maud Committee, but like most of the residents of other counties who had suffered the same ignominy, Harry identified with the old days, with little time for the new administration. Ask the people of Pembroke. Who amongst them wants to live in Dyfed? I am much in sympathy with those who strive for continuity. Even in my lifetime there has been so much tinkering about with our lives by people living on the taxpayers and the rate payers, by those who do not really think the problems through: schooling methods and schools, money, measurements, and vocal minority morals. The polite majority is tolerant, scarcely tut-tutting as architects build accommodation not fit for battery chickens, and multistores tempt the very young and the very old, under threat of prosecution, with goods they cannot afford to buy. The professions have been replaced by career structures and professional managers of everyone, who take pride in being able to forget the basic tenets of professional and practical thought and effort, as everyone is exhorted to elbow out his fellowman, as it is his own duty to "do well," meaning to talk about doing something while doing nothing except expressing his opinion. There, I'm doing it myself! Advisors never make mistakes, because they never really do anything. Beware, therefore, before you take the advice of someone. First take advice on the advisor, and then double check with another advisor.

Anyway, Harry loved "the Middlesex," though, like me, he drew his pay from Surrey.

When I travelled up from Guildford to see assistant county engineer

Eric West and Martin Humphrey about the county improvement scheme they wanted me to control, I went to the county hall in Kingston. Martin, who had designed the scheme, gave me a set of the plans, and we then visited the site, where there was short-term accommodation available in the Middlesex highways compound, now half-heartedly made available for use near the hospital on the A30, a casualty of rationalisation.

While waiting for Mowlems to start work, there was a period of perhaps a fortnight when Harry and I were able to make a comprehensive photograph coverage of the area including every property adjacent the urban route, which existed as a single carriageway having a tremendous camber from the centre down to the sides, an old, well-used, strong-based road, but ready for a facelift. The new works would incorporate the existing road, with surface scarifying and reshaping, as well as new footpath construction, married in neatly to all frontagers' properties; also a new carriageway would be constructed on a wide unused width of land on the north side adjacent the Staines Reservoir. The existing crossroads in the vicinity of the hospital and the Bulldog Hotel would be modified and furnished with sophisticated new traffic lights, automatically operated, to control the traffic and pedestrian ways. There would be tailing in at the ends to marry in with the existing dual carriageway at the London end of the trunk road and a single carriageway at the Staines end, for good measure, diversions of services and deep drainage. In short, it was just a routine job of work requiring attention to detail, kind of refresher course for me.

Christine and Lin soon began to settle in at Guildford, where we had moved in August, 1971, at the time I began to take an interest in the Bulldog Scheme. Lin was active at the cathedral, and Christine landed another secretarial job in a factory in Merrow. As we were not far away from Woking, my daughter's friends there were still accessible, and she also made new friends local to our new home. She was soon to realise her wish to become a student nurse, however, when she was admitted in that capacity in St. Luke's Hospital in Guildford. It has always been Christine's delight to have loads of friends, and the hospital environment suited her fine.

Lin's aspirations to have her daughter marry a Welshman gradually faded as more males appeared in Christine's social circle, and with former friends so far away in South Wales it was likely, for geographical reasons, she would assimilate with Surrey people. At twenty-four she had had numerous pals in a number of young peoples circles, and quite a few of them will remain as lifelong relationships, though as we all know, you cannot see all of the people all of the time. I like to think the twenty or more male platonic friendships had been invaluable in opening Christine's mind to the wonderful variety in human nature. When 1972 came in, she had met a friend of one of her colleagues at the hospital. His name was David, like our second son, though he was nearer to Christine's own age than that of our boys, Newton and David, who by now were both married and living in Africa and the USA. This friendship developed quickly into something permanent, and it was soon apparent that Christine's nursing career was going to be shortlived or at least interrupted. One day, David asked me for Christine's hand. It was a courteous Victorian request, and I had no reason to refuse, my daughter being long past the age of consent. David was also starting out in life as a civil engineer.

The marriage of Christine and David took place on August 26, at Busbridge Parish Church in Godalming, David's family home, and I was pleased to have my own brother, Horace, his wife, Betty, my sister and brother-in-law, Iris and Ron, and my other sister, Thirza, all down from North Wales for the occasion. From South Wales came Lin and Christine's friends, the whole occasion being uniquely enjoyable for Lin and me, as it would be the first and last time we would attend the wedding of an offspring, though we had had four children and had been married thirty-three years. I must blame my roving life for our having a far-flung family in our old age. At the time we were on a high, though it would have been nice to have had our two sons present for the occasion. The two lovebirds only had eyes for each other, so we could hardly ask for more.

Where did they go for their honeymoon? Would you believe, on a chartered six-berth yacht sailing round the Channel Islands with two other couples? David has a passion for sailing. I had retained some *sympatica* for seafarers, but the sailing bug had long released its hold on me after my grammar school holidays and life with my Taid Bithell of Flint on the *pilot*.

August 1972 signalled another landmark, the final release of the contractor's maintenance obligations on M3. Adrian Vickers, the contractor's quantity surveyor and representative, cleared up the sundries with reasonable despatch.

Stan. Marshall, Bovis's quantity surveyor newly appointed to prepare the contractor's claims, had not been wasting his time since the opening to traffic, and now a pile of claims documents, all drawn up and complete with costs being sought, lay with S.E. Brian Oldridge in the office of the superintending engineer at Guildford. The S.E. gave me a ring at Ashford, and we agreed he should send some up for my perusal while I was finishing off my work at the dualling scheme, due to be completed in the next two months. Up at Ashford, the job had arrived at the stage of the carriageway final surfacing. Much switching about of traffic was entailed to enable the works of a bitty nature to proceed. Eventually, having tried the whole volume of traffic on the new twenty-four-foot-wide carriageway one Friday and been submerged by the volume, despite all police efforts, the contractor found it necessary to allow the old carriageway to continue to take the flow from London to the south while carpeting the surface, one lane width at a time, in asphalt. A short length of urban dualling job can be more complex than a few miles of new trunk road across open country, just as the pipe connections for the Comdonkin Service Reservoir in Swansea (vide chapter 12) were more complicated than some parts of the trunk water mains. However, all things come to an end, and the day came when the sophisticated traffic control lights at the Bulldog crossroads replaced the old lights and traffic flowed on the A30 as never before. It was just a short length in a programme of general improvements along the main road from Land's End to London. Take a look at the map. This piece was the other extreme of the A30 from Land's End, at the threshold of the metropolis.

Mid-October 1972 saw me coming to grips with the substantial claims already in to cover the M3 contractor for the particular circumstances under which the work had been executed. These claims made interesting reading, and I found myself having to assess the whole conduct of the operations from start to finish. Stan. Marshall was an artist at preparing claims, added to which he had obviously kissed the Blarney Stone. He rang me up in the Guildford office as I sat at a large table reading through the claims, anxious to get down to the discussions. After about three months of dedicated application during which I burned a lot of midnight oil at home, I was able to submit, to the proper channels, my assessment of what seemed to me to be clearly admissible elements of the claims, prior to the discussions commencing. After a similar period for assimilation, the proper channels were satisfied that I should now take up my role as the front man in the negotiations.

I felt like a quasi-arbiter, having to take each line by painstaking line of each constituent part of each claim, as Marshall and I explained our points of view and costings, discussed items, argued, and came to provisional terms acceptable to the parties to the contract, in the light of all available detail. "I have done so much personal arithmetic." Those were the words of a one-time eminent engineer, Harry S. Waters, when acting as an arbiter for the settlement of claims made by my father for a contract he had carried out for the extension of the Congleton Sewage Works in the early thirties when I attended the arbitration hearing as a boy of eighteen. Well, if you want to get down to the facts of the matter, you have to do "a lot of personal arithmetic." The M3 was a good deal bigger than the old sewage job, but that provided a good insight to a teenager, seeing our agent, Fred Robbins, and my father presenting the case of the contractor, while Edward Wilson Dixon and his son Ted, the consulting engineers, presented their facts and figures before Harry Waters, the arbitrator.

In working up the figures on our side, I was assisted by Vic Bridges, the M3 Q.S., and Nelson Priest, while Roland, my deputy, pressed on with a multitude of rates with Adrian Vickers. When it came to negotiating claims across the table, Adrian and Roland were invariably in attendance in support of Marshall and me, respectively. From time to time in came John Downs, the R.C.U. senior quantity surveyor, to keep the director posted on progress being made on settling the claims. It was round the turn of the year 1973/4 when the parties to the contract were satisfied with the negotiated settlements of the claims presented.

In the meantime, Stan Marshall had had his name entered in the *Guinness Book of Records* for "holing in one" twice during a four-ball (or something of the kind). Of course, to a man with Stanley Marshall's Irish nationality, the case of Guinness that went with the signal record may have been more appreciated than the Guinness entry.

Lin and I had a two weeks' trip around France at the end of October 1972, just after I had seen the end of the Bulldog site. We had travelled via Le Havre, through Le Mans, Pompidou's Bordeaux, and Biarritz. We were almost able to contact our son in Nigeria by telephone from Biarritz, but the delay in getting through made it not worth the trouble, so off we went along the north side of the snow-topped Pyrenees. Lin started to display a star-shaped pattern on her right cheek after an insect bite, which was getting progressively worse until, when we arrived at Pau, we were compelled to seek help. We ascended the cool, clean stairway of the town hospital. At the top of the stairs we became aware of a long corridor with various doors leading off on either side. A nurse approached. We chatted in French, her fluency tolerant of my best Anglicised French, the subject being obvious. The young lady explained that the surgeon would soon be out of the theatre and would look at Lin. Pretty soon out he came in a white coat. He had spent some time in England. He looked carefully at the cheek; then we went into a reception room on the right, the surgeon, the nurse, Lin, and I. While our friendly doctor chatted away, his assistant was preparing a stiff white paste. After a while it was ready and the doctor spread it on Lin's starred face.

"It will take a few days but it will come off clean," he assured us.

We were relieved and I asked, "How much, Doctor?"

There was no charge. *Vive l'entente cordiale,* I thought as Lin and I expressed our gratitude and off we went.

By the time we got to Perpignan, the face was cleared up. It was the last day of the summer season, October 31. We bought a large, heavy ivory vase with carved figures all around it and a coloured metallic fruit container for the sideboard.

After we made our purchases, the shop lady closed the door and came out with us. "C'est finis, madame, m'sieu," she said. She had closed till next season.

Returning through Narbonne en route for Toulouse, we were given bunches of grapes when stopping at the roadside to see the vineyards, and at Toulouse we picked up souvenirs to remind us of the *concorde* and the violets.

I thought the whole of our fortnight was to be trouble-free apart from Lin's face until, as we approached Le Havre, my Renault 16TL conked out! A red light flashed on the dash. Some youths gave us a push start, and, as we had some time to wait for the boat, I decided to push on to Dieppe, where my car brochures indicated a Renault service garage existed. We toddled along making sure always to stop on a hill. The garage replaced the belt driving the dynamo, mine having disappeared when it broke, hence the bump under the bonnet and the red warning light. It was also necessary to charge the battery because that was flat. On the way back to Le Havre, along the coast road, we stopped a while at Veules-les-Roses, enjoying a lovely lunch. As we continued on the final stretch of our journey, we were struck by the multicoloured Brittany cows, colourful to an extent I have not seen in our own British cows. We caught our ferry back to Southampton and headed for home. It had made up for the foreshortened holiday we had had in 1968.

During the fifteen months I had spent working on the settlement of the M3 contractor's claims, November 1972 to January 1974, I had been toughening up physically quite appreciably. This was because I had to walk up and down Guildford High Street, with its one-in-seven slope, as the subunit had offices at both ends and my work associates and records were accommodated in both. The documents were heavy and bulky, but they were nicely accommodated, a few at a time in a well-used leather briefcase of mine, referred to by Alistair Foot on the M1, when he first saw it as "a briefcase and a half." The heavy document case, plus the walks, plus the running up and down steps at either end was a "kill or cure" routine, which I happily survived despite my advancing into middle age.

In the Guildford design offices, the drawing office was putting the finishing touches on the Esher Bypass scheme and meetings were being held with people from public utility services and others to include as much detail as possible in the documents. I was able to read up the back history of the scheme in preparation, a route through the "Stockbroker Belt," always expected to be difficult from the point of view of having to satisfy resident associations and individual owners of affected properties. Alan Brown, whose preparation work on the scheme was drawing to a close, was shortly to be transferred to county engineer headquarters, then at Ewell, where he would become assistant county engineer. Senior engineers Dave Rowe and Colin Batterbury pressed on with liaison and design until the documents were completely ready and tenders invited, in due time being received for examination and analysis.

A civil engineering contract of this size is not just a simple bid to do it all for a sum of money, despite requests for unqualified tenders. Letters usually flow in during the tendering period for clarification. These have to be answered and both question and answer circulated to all tenderers so that the bids may be made on the same basis. Even then, with so many thousands of details appearing on drawings and other documents and the explanatory correspondence, there still remain questions of arithmetical accuracy to be cleared up. The whole tender must be seen to have integrity. Everything must be seen to gell, and verbal explanations of what a contractor understands as his reading of his obligations will not have value unless these have been confirmed in written form similar to the signed contract. Meetings to wrestle with ambiguities, contract starting date, and all doubts must occur in order to get the work started on the right footing. One should never proceed on the basis of, "Get on with it and we'll sort out the teething troubles as we go along," or some such trusting basis. Certainty is what is to be aimed at so far as the contract premises are concerned. Eventually, after Dorking Headquarters meetings, it was decided that the starting date for the Esher Bypass construction, seven and a half miles long and costing £11 million in 1974 terms, would be July 1, 1974, On that day, I was called upon to attend an interview for the job of C.R.E. for which I was shortlisted. As it turned out, I was accepted about midday and went straight to see the site of the works, which had been let to W. and C. French Ltd.

Although I never dreamed of such a possibility at the time, the final drama of my work at Surrey was about to be enacted, the final contract I would supervise in Surrey County, before retirement.

Ken Sear, now on seat as superintending engineer at the Guildford Subunit offices, had earmarked John Jackman to blaze a trail on the site with the accurate establishments of primary setting out. John was already in the field fixing those intersection points for changes of direction, those tangent points for starting and finishing of the curves. He was doing for the Esher Bypass what young Price had done in 1940 for the depot road at Trecwn Naval Armament Depot (vide chapter 6) and what Keith Booth had done for the Burton Bypass, (vide chapter 16), and what Henry Criswell had done for the Great North Road. Let there be no mistake; without the setting out engineer a job does not start, but he seldom merits a mention unless he makes a mistake. John was unobtrusively setting up his control points in the field, and Tony Knight, an R.E., was esconced in a field caravan office alongside Copsem Lane, Esher, when I came in on the afternoon of July 1, 1974, to make my number with Tony and talk to one or two of the men who are always thrown in to make a mark for the contractor on the official starting day on the site.

Over in the contractors' caravan a very tall, straight personage presided for W. and C. French. It was Nigel Pilgrim. Nigel was a character and a half, standing as a sort of John the Baptist in relation to Esher Bypass, rather like Brian Kearsey of Costain did to the Usk Reservoir scheme pipeline for Swansea (chapter 12). Pilgrim had to get things off the ground for the contractors. He had to provide instant know-how on everything, as he was beleaguered by queries from locals about the contractors' vehicles entering the hallowed grounds of the Surrey woods in the heart of the Stockbroker Belt. A host of people milled around with every intention of safeguarding their traditional rights. Nigel had to receive all comers as courteously as possible and give them assurances where they would be affected by his firm's intended activities, handing them over to Tony when queries related purely to the concepts of the scheme as laid down by the authority.

I collected the few unanswered letters, queries, and bits of paper from Tony so they could be dealt with in the office on my return to base at Guildford, while work in the field began to be organised by the contractors in the "Pilgrim Way."

There was a period of time of about three months, when the sun shone on the righteous. The long awaited Esher Bypass was getting under way and really favourable weekly progress reports coming in, and the innocuous nature of what few complaints there were, together with an orderly development of contract business in the paper engineering world of the executive and administrative offices at Guildford and Dorking, gave one a feeling that all concerned were enjoying the experience.

Bill Kirkland had joined me as deputy and was to prove to be a backbone to the organisation in his dealings with outside bodies, the public utilities services and the police, one or two of whom he had worked with when doing some work in the Greater London area prior to his arrival in Surrey as a potential R.E. in the early days of the M3 contract work. Since then Bill had put in time as deputy to the C.R.E. on the second M3 contract, now under way. Bill was nearer his home at Esher, which was convenient for him and a windfall for me. Bill was a man who could work on his own, get something done, and tell you about it.

What about that man in the thick of it all in the field? Nigel was rolling his plant through newly cleared forest areas, burning the scrub, making a track across hill and rill, with pipe culverts to get him across, like an army captain advancing on behalf of Bill Slim in Burma. Nigel was claiming the site, and how! In those first few weeks, no more than three months, Nigel had claimed the site with fencing and had a form of track from end to end of the route of the new seven miles of motorway standard highway to relieve the town of Esher and ease the chagrin of the motoring public travelling up the A3 by the end of 1977, December 31 being the date for completion.

Everyone could now do an inspection up and down the site in a Landrover. Some even used their cars, though the sandy surface of the track could be dusty. From time to time Nigel would be seen waving imperiously as people tried to cross awkward points where the sandy gravel road access track had suffered overly in use by drivers eager to use the "Pilgrim Way."

Back in the Guildford office, Kathy, who had been on the site at Camberley, was my new office manager. We had been allocated a room to get the paperwork off the ground. Files always have to be set up to accord with the manner in which the contracts have been prepared, conditions, specifications, drawings, bills of quantities, and related subjects, owing to contact with public and private bodies. There has to be stationery acquisition for formal correspondence, site orders, variation orders, and records of all sorts. I had to be looking in to the sites of the works during this initial period as frequently as I would when firmly established in my C.R.E. site offices as soon as they were available. While these were still being erected, I would invariably make a Saturday morning tour of the site, picking up a clerk of works, Dave Budd, ever ready to keep things in good order insofar as his own section of the work was concerned. Some thought Dave a bit abrasive. I found him enthusiastic, observant, and incisive but fair.

What is it about roadworks that makes the gods play games? In 1958, we on the M1 (chapter 14) experienced a hell of a winter, with rain well over average. Each week, poor John Michie, the chief project manager, sank progressively lower in his chair at the Friday progress meetings. Here at Esher, when all seemed set for a fair wind and a quick sail on a happy voyage to complete the journey on calm seas, we began to experience the dubious delight of exceptional rainfall. The site was claimed. We could have stood a bit of rain, but not that torrential stuff. For a time there was optimism as work continued at a slower pace to accord with the season of autumn. Then the persistent pressure of precipitation caused the bogging down in the field so familiar to winter earthmovers. Site visitors could not get through to our offices to make their business calls without muddying their cars. Pilgrim was obliged to think of stabilising the long access through the woods necessary to seclude the works compound, as he bore the brunt of criticism from hordes of uptight superior sophisticates, having the benefit of hindsight but who had also not come up with any timely clairvoyant suggestions in the preceding weeks.

Perfidious circumstances of oversandy content of the granular excavation (Nigel had taken the poor stuff to keep the best for the permanent works) and over saturation caused caustic comments from folks wanting to know why the sandy tracks and surfaces had not been "sprayed and chipped" before the winter came in. (My mother used to say, "Water is a good friend but a bad enemy.") It was necessary to do wholesale concreting of the access to the offices before a satisfactory state of affairs was achieved. Not so in the field, however. It continued to rain heavily throughout the winter, with about a year's precipitation in four months, so we would be very lucky to attain outputs akin to even the usual reduced production for winter working, normally about a quarter of the summer levels, with activities limited to moving of services, piling and foundations adjacent to existing roads, and tiny earthworks in local accessible places.

We had expected lots of difficulties with residents during the construction of the Esher Bypass. Those who had been charged with the preparation of the contract had experienced alarm and despondency from those to be directly affected by the inevitable disturbance of their privacy, to say the least. The public inquiry had laid many ghosts, but the fact of the road being built remained a fact to stick in the gullets of the local people. Besides under the scheme, there was the intention to import chalk wholesale from the direction of Leatherhead to make up the deficiency of material available in the cuttings to complete the embankments.

Now in the event, I was surprised at the great courtesy we experienced from the local people. We conducted their representatives along the site to show what we were doing, and many were taken up with our efforts, so that despite some grievous anxieties, we were not assailed with threats amounting to injunction that I can remember.

W. Dyer came in as project manager for the contractors and the complaints of the piling vibrations were discussed in a friendly manner with the residents affected. Some such complaints introduce grey areas of responsibility in contracts, and insurers can be the most careful of people when asked to pay out in circumstances that are not 100 percent their insured's liability. The result is that even the most patient casualty of the works can become demonstrative during the process of time. sometimes, for such a one, action speaks louder than words.

One day, while we executive types were all chatting away at a monthly progress meeting in the site conference room, there was a thud against the door. As we were absorbed by our mainly self-satisfied observations as to progress, we did not even bother to go to the door. John Melrose, the current county engineer, was in the chair. Our meeting dispersed and when I returned to my office I learned that a barrow of muddy sand had been tipped against the door by our nearest neighbour, the object being to draw attention to the inconvenience he was suffering in the form of a muddy surface in front of his immaculate country home in Copsem Lane. The impact intended for the meeting had been lost, however. Vic, Q.S. now assisting Derek Scrase, S.Q.S., had taken swift and silent action to shovel up the sand and remove it, so that not a grain remained. When our residential friend and neighbour learned of the sequel to his excursion, he could but laugh.

Len Brown had now taken over the contract as project manager for W. and C. French Ltd. and was getting to grips with the work, showing every intention of completing the outstanding works within the contract time. Len was direct and easy to work with—another Wally Barrett (chapter 16). Len was consumed with the practicalities of the project. We did not have on site the amount of coarse, sandy gravel deduced from the boreholes. If Len could win more gravelly soil from private fields adjacent to the bypass, it would serve to reduce or eliminate the need to import all that sticky chalk that some of the local residents feared so much. It would also mean a very useful source of supply of natural gravel, useful for drainage, filter drains in cuttings and in layering embankments with granular strengthening material, for concrete aggregates, and even for asphalt layers in the carriage construction. About a mile length of the earthworks' method of construction was reviewed, and in simple vein, Len's proposal for a sound reappraisal was accepted. Good for the contractor, who had his head with working the difficult length, good for his client, the secretary of state for the environment who was safeguarded against additional expense inevitable from the type of ground to be encountered in the wet cuttings.

Len Brown had his teeth into the job and began to make up for time lost prior to his arrival on the site during that dreadful first winter, though there had also been a number of changes to the work justifying an extension of time amounting to two and a half months beyond the original date. Early in the final year—1976—I had jokingly said to Len, "If you finish this job in the original time, I'll give you a gallon of ale"—just a remark en passant at a time of seemingly insuperable difficulties timewise.

During that final year the progress graph began to point up and up. The pressure was on and the battle was being waged to complete by the revised date in March 1977, with prospects looking good. Bill Pomphrey still hovered around, the stalwart project engineer liaising between the sites and the Dorking office. When it became clear that there was a sporting chance of achieving the original completion date, there was a great stir among all concerned parties, as coordinating meetings heated up.

Road surfacing (the asphalt plant was actually specially erected at the site for exclusive use of the bypass work), road signs, lines, road studs (cats' eyes), soiling, seeding, stressing up concrete bridge docks and erecting parapet rails and safety fencing, roundabout lighting, and a host of operations all seemed to be proceeding at once in the last burst of life that gave birth to the bypass, in which so many people performed as midwives.

Bill Pomphrey was now in a quandary once again. Not only did he have to conform to the requirement to give two weeks' published statutory notice of the opening, but there was also the problem of the Christmas holidays immediately prior to the original date. It was not going to be on to have the road opened on New Year's Eve, slap in the middle of the holidays. Now all we midwives were well aware of the long-term forecast for snow, so that the road would have to open before Christmas or it could well be much later due to weather considerations. Len called in reinforcements from other jobs farther afield. In the end it was decided to open on the fifteenth of December 1976. That was two weeks ahead of the original contract date and three months ahead of the revised programme with entitlements for extension.

At half past five in the morning of the fifteen of December, I met Bill Kirkland, my deputy, at the Painshill end of the site. Nobody envies a deputy. He is the man who has a welter of detail to pick up and saw, each one larger than life. Bill had been breathing fire and brimstone for weeks, even months, to chivvy people along, clearing out drains, getting the work up to scratch. Now he was looking for perfection with just a few hours to opening. The day before, the site had looked dirty. Overnight Len had introduced the bowsers and sweepers and brushes. Teams of people coming up in the rear were like so many Mrs. Mops, all to present a new, spic-and-span appearance.

After a run-through, I could see the opening fixed for late morning was going to happen. Bill's batteries were still highly charged, and off he went to liaise with the police and others at the difficult traffic switching operation where the new route would merge with the old at the London end, at Hook.

Well, the Esher Bypass did open and brought with it a generally favourable public reaction. Len deserved his gallon of ale, which I only just managed to present to him at a site assembly to enable all to take some satisfaction, the contractors and the engineers' staffs, for a good effort all round. I say only just, because as I came to hand over the prize, it was not there. Bill, my deputy, always the humourist, had made the tin disappear, then produced it, Paul Daniels fashion, at the last second. Len's cup of cheer was a trifle but spoke volumes for the staff's combined effort and our tribute to the contractor responding to the challenges of the bypass.

As for my own staff, they had been second to none in my experience. The thousands of items to be monitored in such a work call for the special talents of the whole spectrum of people necessary to form a balanced team. The achievements, separately attained, make up the whole. It is one for all and all for one, a service to the community.

The M3 at Camberley had been initially split into two halves, with each section resident engineer having a mix of work on the roads and the structures but with the surfacing works coming in at the later stage as a third section. The Esher job was operated from the beginning with two R.E.'s Tony Knight and Cliff Brown having the day-to-day supervision of the roads section and the structures section, respectively, with the backup support of the materials engineer, Terry Howells, each section having a suitable number of junior staff.

Derek Scrase, assisted by colleagues, held his end up on the quantity surveying section so that no insuperable problems remained in that direction by the end of the work, though there are always a number of accommodation works and sundries to be carried out during the period of maintenance after a highway has been newly finished and opened to traffic.

Our administrative work under Kath gave the vital support that is necessary to enable everyone on the staff to do their own jobs effectively and efficiently.

One interesting aspect was the early days' archaeological explorations that took place on the Crown Agents land, where someone looking remarkably like Margaret Rule of "Mary Rose" fame was in charge of the people she had scratching about on the site of a former Roman camp, various bits of old pottery of the Roman era being unearthed and salvaged prior to the onslaught of the excavators.

The onset of the recession, with its effect on public spending, upset the lives of so many people. I had been thinking that my permanent status in Surrey, after a lifetime of moving around the world, gathering no moss, would mean job security until I was sixty-five.

Lin and I had spotted some plots on white land for sale at Farnham and had bought one in the belief that by the time we retired we would build a little home to our own ideal requirements and make out, with some dignity, for the rest of our natural lives. It was not to be. Before Esher was completed, I inquired of the county engineer what would be happening to the roadworks programme in general and me in particular. It was made clear that county authorities were already under pressure to shed staff. My completion of the Esher Bypass would lead to a "cul-de-sac." Staff who could be declared redundant, and I came under that category, would be treated humanely, at any rate, by comparison with the time I was superfluous to Admiralty requirements in 1947, when I felt I had been fortunate to leave with a job to go to and not one penny beyond the salary paid to my day of departure. The latter was the world I knew, and my treatment by Surrey in the modern world was much appreciated, even if it changed my domestic plans. there were millions worse off than me, and I am not one to complain if everything does not come my way.

Lin and I managed to earmark a small, modern house in South Wales, and 1976 saw us visiting the property to get the gardens ready for occupation when I would be no longer essential to the road programme. Eventually it was agreed that I would leave on my sixtieth birthday, the sixth of February 1977.

One day Lin and I set off about half past four in the morning to travel to Barry, and, by a quarter to six we were sailing along the M4, about ten miles west of Membury Service Station, when our white Renault 16 TL was struck from behind. I thought it was the failure of a welding repair to the underfloor of the body work, done on the previous day for my M.O.T. test. Instead it was a car coming up fast in the mist. My car was a writeoff and the story too sordid to tell, but it meant that I finished up without a car and Lin without one of two matching pouffees we were taking down to sit on as we used the house as a base while preparing the garden. The pouffee, which was stolen from a wrecker's yard to which we were towed in Swindon, is still missed by Lin. The car was replaced under insurance with one equally old but good enough to see me through the envisaged span of my working career.

It was during our hard week-end gardening efforts that I lost something else—a muscle in my abdominal region. As I returned from walking to the mile-distant Weycock Cross Garage to get paraffin to warm the garage at the back end of 1976, I felt a pain in the groin. Suddenly I was immobile and could not lift the can. I put it down and rested. Lin

Motorway in Surrey.

patiently carried the can while I struggled slowly back to the house in acute pain. I thought I had a cramp. It had been a busy day from early morning, excavating in the boulder clay soil around the new prefab garage. I had been overdoing it—a semisedentary type thinking he was a navvy.

Back in Guildford the doctor diagnosed a hernia. I would have to go into hospital to have it sewn up. It was the beginning of October and I wanted to see the opening of the bypass. I decided to wait until the realisation of that event and managed to go easy on the walking, which was limited to about two hundred yards at one stint; otherwise immobility would set in. Softly, softly.

Bad luck does not come singly. The recession was hitting the housing market, and we were having a job to dispose of our house in Guildford, which was a must, as we could not afford to keep up two houses for very long. The price had to be dropped considerably as we moved into a buyers' market. We were let down by a purchaser who did not complete on the day we moved our furniture to our Barry house, despite all arrangements for the occupation to suit the purchaser, so we had an empty house on our hands in Surrey for an inordinate length of time. In the end, with the return of my daughter and son-in-law from Africa, they arranged to take it over at an agreed upon price in the latter end of 1977.

When it came to leaving Guildford, Lin was not as pleased as we expected. When you go to live in a new area, particularly late in your life, it takes time to build up relationships with new friends. Lin had gradually been identifying herself with several good friends in the area and at the cathedral only half a mile away. She was almost ready to break down and cry when a kindly letter came on behalf of her colleagues thanking her for her efforts at the cathedral. Now realising what our departure would mean, on the debit side, we both had mixed feelings. I would have a lot of life and work to look back on, but I would not be going to another job. I was being "put out to grass," and that is what I will be telling you about in the next chapter, when I have explained how this episode finished up.

On January 27, 1977, a large assembly of colleagues and friends of the county highways and road construction unit gathered in the conference room of the Esher Bypass to pay me the honour of a sendoff. John Melrose, the county engineer, was generous in his preroration of my major highway and motorway efforts for my country. He did not miss a detail as he ran through my career on roads, with an inside knowledge far more succinctly put then I have portrayed in the episodic diatribes in this book. It seemed that during my progress from job to job John Melrose and I had crossed paths on more than one occasion.

I took it as a great compliment to see so many colleagues with whom I had been closely associated gathered round in a thickly packed arc in the conference room. There were friends from all over the county whose valued signatures will help us recapture their faces, one by one. I was overwhelmed by the turnup of staff members from John Melrose's county engineer offices, Brian Edbrooke's Dorking R.C.U. offices, Ken Sear's subunit offices, and Neil Trickett and Don Cox's C.R.E. offices at Thorpe and Merstham, fellow personnel, continuing the good work of road building. As I looked round their faces, the ones with whom I had shared experiences with over the last eight years, I could see behind them others, stretching back like a cavalcade, right to the time when my father first built his own bungalow, with a little help from his friends, in the very early twenties. But the words I was moved to speak related to the type of work we had all shared in Surrey and those on roads, everywhere, my poem on:

Roads by Price: (*Top*) depot road, Trecwn, in 1972, after thirty years of use; (*middle*) stretch of M1 in 1980, after twenty-one years of use; (*bottom*) Via Nova Plant at site of Esher Bypass, opened in December 1976.

Roads

These are not things that royalty grace.
They are so common in their place.
In all their shapes and sizes
They have their own grace.

Sporting studs and paintmarks,
Signals, posts, and signs,
Bearing our essentials,
They are our lifelines.

Strength builds roads,
Roads build strength,
Used by all,
Length after length.

Crooked and straight,
Level and steep,
Over the hills,
Unto the deep.

Older and younger,
Keeping anew,
Geography, history,
Heritage true.

And so to retirement. . . .

19.   Out to Grass

When I first came to Barry to retire on my sixtieth birthday, I had a new challenge, different from all my previous ones. I had to learn to "chew the cud" without getting on Lin's nerves. She had never had me under her feet at home as a permanent arrangement before. She too would have to live with a surfeit of my company she had not known in nearly thirty-six years of marriage.

Now six years later, I put pen to paper to write my retirement episode, we are still not bored with each other, so we must have found a way to carry on reasonably satisfactorily, whether by accident or design I cannot say. Perhaps through my relating the main events as I recall them, you as my reader may be more discerning as to why we are reasonably contented with our lot. We do have a different outlook on some things, but maybe as one grows older one realises that one has a lot more things in common.

For one thing, I had an innate sense of thrift and budgeting for reasonable financial security in old age. Against this, when one travels around to find work the outgoings are heavy, and therefore I had not been able to amass so much in the way of money. As my pension would not be great, less than a quarter of my salary when working, I had to make the move to a cheaper retirement home, both as to capital cost and outgoings, in order that I might have a little nest egg for a rainy day. I know people say "invest in property," but there comes a time when one has to be realistic and really rationalise income with expenditure. A big, expensive house is no use if it means the worse of the two Micawber situations. So we had bought a small house in an estate being newly built. My physical overwork preparing the garden had already caused a hernia diagnosed by my Guildford doctor during October. We transferred our family medical cards to Barry, and I did not waste any time in having my new doctor, Dr. Parr, satisfy himself on the hernia situation. My preadmission examination was arranged at the Amy Evans Hospital. It came about that my admission to Llandough Hospital was to be Friday, the eighteenth of February 1977, on which day I was X-rayed and allowed to stay at home for two nights, returning on Sunday for the Monday repair to my anatomy. In the ward of some twenty beds, there were a few cases similar to mine, men of all ages. The time for my operation was fixed for 2:00 P.M. Monday, and at that time my body lay prostrate on the trolley in the corridor outside the door of the operating theatre, prompt to time. Bill Kirkland had had experiences of operations and had explained that, "Nowadays there is just a prick in the arm or the leg and you are out to the world."

One moment I was prostrate on the trolley in the corridor. The next I was vaguely aware of being rolled from the trolley onto my high hospital bed back in the ward. Then there were two hazy figures sitting by the bed. Lin and Christine had come to see me.

"I have to go into theatre at two," I insisted several times, with consciousness fitfully returning.

They did not argue with me. "Just rest, Jim. It's all over," Lin said once.

"You've been in, Dad," said Christine. "It is six o'clock."

Still I was insisting, "No. I have to go in."

It was good that Christine had come down to help us with driving the car during this awkward time, and she brought Lin in each day for a few days of recovery. The care one receives in hospital is something special. There are always worse cases than you around. One old man had a drip on. A younger man groaned in agony for days after his hernia op. He had ruptured himself using a rotovator in the garden. Another man, well in his seventies, made light of his triple hernia, taking the tea wagon around for the staff while waiting to go in to the theatre.

The white cast-iron rail, too high off the floor, was a brute. More than once as I gradually rose to go to the bathroom, I sat on my swollen right spherical and felt the ultimate in pain, which served to restore my sense of caution.

"Don't cross your legs," ordered the staff nurse as she strutted down the long aisle between the beds in the ward. If she said it once, she said it a dozen times always with the same care for the patient. After a couple of days, a group of students were on tour with the house doctor. The mixed group looked at the specimen between my legs, with the size and colour of a new red leather cricket ball. "It will settle down," was the doctor's note of assurance to his class in answer to a question, which was nice for me to know, too, though I still had my doubts.

After an operation, bodily functions can be a problem. The curtains were drawn as the thunder box was brought in and the enema produced by a sympathetic nurse after I had had no result for an undue period. I had no more trouble after that. Finally, as I was sufficiently mobile for discharge, my stitches were removed. What made me think that process was to be dreaded? The angel who removed them did so with such delicacy they seemed like spiders' threads—no feeling at all.

Postoperation care was advised. "Take it easy," everyone said. "No gardening, Mr. Price," Vic Bridges had written on March 22 from his office desk in Surrey. One outcome of my period of hospitalisation—before and after—was the writing of twenty poems. It took till May 31 before my doctor declared me fit to resume my retirement or work availability, during which time only minimal physical effort was possible. Soon after that date, however, tiny areas of concrete started to appear to extend the paving to patio and paths around the rear of the house until, after a few weeks, I had the first pangs of guilt at not being a member of the working public. A long vacation is one thing, but this was ridiculous. How long would my funds last? Was I just an old horse out to pasture?

There followed a period of seven months when I pressured the professional and executive register to fix me up in something. Indeed I had been on the register at Cardiff since I first knew I was to come to South Wales in a redundant paid-up, retired capacity.

"You'll be lucky," intimated young Mr. H——. "Not at your age," he expanded.

I began to feel like Jesus—"despised and rejected of men." I managed to have three interviews in those guilty seven months, each for an R.E. supervising an honest job of works construction. There was some road and bridgeworks—a new interchange to be built at Queensferry in north Wales. There were some waterworks, the training of the river Mole, a tributary of the river Thames. Each of these were not on with the consultants as I wished to return home for the week-ends, and each a distance of 150 miles from home. Lin and I were thinking of taking a flat near the work, but felt we were at the time of life when we deserved at least the week-ends in our own place. The R.E. is expected to be on call at all times, at no extra cost to the employer, of course!

In the seven simmering months following May 31, 1977, I had as my two main objectives the release of our house at Guildford, which was empty and could not be occupied until our daughter and son-in-law, Christine and David, returned from overseas, and the pursuit of something worthwhile for yours truly. Lin was more resourceful than I in filling in the time, having had a lifetime of experience in being home-based and pursuing various hobbies, attending women's circles and suchlike.

I decided to buy a new, smaller car to replace the old banger I had bought when my Renault 16 had been written off on the M4. So August saw me picking up an 1100cc type Renault 6, which we fully expected to last us for the foreseeable future of our retirement.

Time passed as we tended the garden, nipping back to cut the hedge and lawns at the Guildford house during the summer months. We also made a trip to North Wales to see my mother and other relatives of mine, my dad having passed away in June 1976.

With the first winter of retirement, we decided to do something positive to break up the winter nights. Night classes were the answer.

The Glamorgan County Education Authority run a big and varied programme of evening classes. At the beginning of term, Monday, September 12, I started to learn Spanish, while on the Thursday evenings I began with Welsh classes. Lin took up cake decoration on the Wednesday evenings. Norman Pereira taught Spanish as if we were already natives of Spain, whence came his predecessors: Bob taught Welsh as if we were natives of Mars, so slow and deliberate was his technique. Of the two, I preferred Bob's way. I was able to read on a few chapters ahead to enable me to appreciate what was on the menu, so to speak, for the evening, all from the textbook. Norman did not recommend a textbook, so it was a world of surprise, but for me, difficult. As it happened, Norman decided to change his night so he could do a foundation course, so I was forced to drop out gracefully, as I was otherwise committed on the Tuesday nights.

One of the joys of night classes with a man like Bob is the relaxed, sociable approach in which the subject is taken on board. Every evening the first thing to happen was for the oldest member in the class—James Price—to report on the notice board news for those who might be interested in any of the opportunities to join in social and cultural activities available through the good auspices of the education organisers. Halfway through the easy, step-by-step lessons, read from the textbook, in we went for coffee time—*Amser coffi*, and during the coffee break we sat as a family for an enjoyable chat in the canteen. Back in the class we pressed on, not leaving anyone behind, and were off at nine o'clock sharp. This technique left those with a real ambition to learn Welsh desperately wanting to press on and, I think, was right for absolute beginners, since I have noticed several of Bob's class gave a creditable account of themselves in Welsh in later years.

Anyway, how's this for a laugh? The consulting engineers for a proposed new pumping station to be sited in Quay Road, Neath, advertised for an R.E. to supervise the work, and I was asked to attend for an interview on Wednesday, December 14, 1977. Before my morning interview, I popped down to look at the site and then crossed over the pedestrian footbridge over the river Neath to view the scene from across the river. Two locals were chatting away in Welsh.

More to make conversation than anything else. I asked them, "Whereabouts is the golf course?"

They broke off their conversation and one answered my question with another: "Can't you speak Welsh?"

I mustered my best efforts from my then still very limited vocabulary. "Well, I'm a Northwalian." (Wel, o'r gogledd yw fi.)

"Oh, Diowl," said the other man. "Speak English. You're too deep for us."

So Bob's teaching had helped me over that hurdle.

By the afternoon's interview, the candidates shortlisted to go before the committee were down to three, with a senior councillor in the chair and others to represent the public, Ken Taylor deputising for the borough engineer, Alan Jenkins, with others representing the officers and with Malcolm Thomas representing the consultancy partnership for the scheme, Thomas, Morgan, and Partners. Each of the applicants had about a forty minute interview, which proceeded much on the same lines as my original interview with Swansea Corporation, as mentioned in chapter 12. In the end, I was called back and offered the job. Neath,-Castell Nedd-is proud of its Welshness, and when the business of the day had been settled, the chairman came over and asked if I spoke Welsh. ("Dychi'n siarad Cymraeg?") I was able to say I could speak a little, as I was in my first year at night school in Barry, ("Do, typin. Blwyddyn cyntaf,yn yr ysgol nos,yn y Barri.") It was a warm introduction. My old Taid of Flint was still guiding me in my grey hairs, up from above.

I was off to a fresh start, and with the blessing of the chairman I had to start in the New Year, 1978. Strange that on the way home my mind should travel back to the anecdote told me by Fred Newall, my mother's cousin and Taid's mate on the *Pilot*, as we skirted Anglesey in general and Amlwch in particular. It seemed that Taid used to put in to Amlwch sometimes and play the organ in the chapel, more from ear than from music." The captain's a *wonderful* player," said the deacon. While the boat was still in the tiny port, then a sleepy little town, the deacon was talking to Taid Bithell, who remarked how nice and quiet Amlwch was. The deacon said, "Captain, it's not always like this, man. You ought to be in Amlwch on Saturday night—crowds of people, man. There's a fair on." The sequel was that the two friends were walking round the fair on the Saturday with only about twenty others at the fair. "Captain, what do you think about Amlwch now? Look at the crowds," the deacon said. The story must have struck a chord in my subconscious as I travelled back to Barry after my interview that night. Maybe it was the enthusiasm for life in a quiet place and the wish to share it with another. Somehow I felt involved in life again in a more meaningful way.

"Nice to be back in harness?" asked Peter, a neighbour, as he saw me toddling off in the mornings to my job in Neath. I agreed it was. Old habits die hard, and after you've been retired or semiretired or made redundant or received early retirement or whatever you like to call it and when you have been acquainting yourself with the fact that you are just as much on call as a man who has still not reached his state pension age as you were when at work, you come to the view that you might just as well be working. However, I had hoped to get myself something nearer to home, but a temporary job forty miles from home was better than nothing, even if it did mean no benefits to me pension-wise and hard going travelwise each day. It could be looked upon as a sort of tapering off job. With hindsight, I must admit that I was fortunate to have that opportunity to do something useful for those two years I was engaged at Neath.

But it proved to be more than a tapering off job. It was a real, live interesting job of work, not something to fade away from my life's engineering association, rather something to let me step back gracefully into the old atmosphere and savour every moment of my good fortune.

The new pumping station site lay at Quay Road alongside the river Neath, behind an old river wall and flanked by old buildings on two sides, a garage and a former pickle factory, now unused. A modern factory premises, the engineering works of John Owens lay on the fourth side of the site. The land had been refilled over the alluvium of the original riverbank, which had seen unrecorded activities in the river's navigating past. It was nobody's business what had been left in the ground by our forefathers. That would not be discovered fully until the whole area had been subjected to the construction works operations. Of course, site investigation—a couple of bore holes and a trial pit—had been performed and deductions made based on what was revealed in the three locations of this comparatively compact site. It would not have been possible to dig up the whole site as a trial hole, so the methods and plans for the scheme had to be sufficiently flexible to be modified to cope with the perfidious vagaries of ground conditions.

In mid-December, before I started work, I took some photos of the commencement of the sheet piling to form the cofferdam, useful photos to bear witness to the situation at that time, the beginning of a story of a construction offering quite a challenge for the size of the site, owing to the obstructive and awkward subterranean conditions. The interlocking steel-piled watertight curtain walls round the huge sump for the new pumphouse were connected to diaphragm walls to make five cell compartments, for reasons of coping in a practical manner with the enormous pressures involved. Work proceeded after the piling, by removing the saturated ground, by grab and dredging with the aid of divers. A concrete plug to seal off the water below the level of the sump floor was concreted below water, using a tremie pipe, and the structure floors, walls, reinforcement, pile cropping, and concreting were done in a logical sequence of operations until the work was up to ground level. The main feature of the sump was the three sloping concrete channels, into which would sit the motorised Archimedian screws, which would literally screw the sewage from a lower to a higher level, so that the sewage of the borough would no longer be discharged into the river but taken by piped sewers to a discharge point out to sea, under another contract, in the scheme of things.

The execution of the works was to give truth to the old saying "No two jobs are alike," and we will go into some aspects of that together a few pages further on.

In the meantime, let me describe, in general terms, life for me in those two years 1978–80 while I was working at Neath.

The journey to work used to take me about fifty minutes, which meant starting out at seven-thirty to arrive at eight-twenty. The way led past the new Rhoose Airport, opened in 1972 by HRH the Duke of Edinburgh and a testimony to the Glamorgan County Authority, whose airport committee chairman was Councillor Leslie Richards, J.P. After passing the airport on my left, the B4265 route bypassed Llantwit Major, which also lay on the left, and then led on to pass the RAF station at St. Athan, which lay on my right. Continuing on through the villages of Wick and Ewenny, one met up with the A48 at a crossroads on the outskirts of Bridgend. There I had to make a left turn into the main through road to the west, but often had the darndest trouble with the engine at this crossroads. It had a habit of conking out! Starting up again involved a lot of persuasion. At first, it seemed like the carburetor was starved of petrol.

At Neath Car Sales, the Renault agents, Ralph had a new petrol cap fitted to make sure there was no vacuum created in the tank. None of the Renault garages could trace the idiosyncracy. The engine cutting out only occurred at that crossroads when going to work, not when returning. In the end I could only conclude that the overhead wires on the side of the road for a distance before arriving at the crossroads had some kind of induction effect on the ignition system. Later I avoided the spot by making an earlier turn left at Ewenny to join the A48 a little farther to the west.

Beyond the town of Bridgend lay Stormy Down and Pyle. If I were low on petrol, instead of bypassing Pyle by joining the M4, I would run on through that town and fill up at the garage just beyond the lights. Afterwards I joined the M4 in time to continue my journey along the Port Talbot Bypass, with the gigantic Margam Steelworks on my left. This always had nostalgic memories for me. (See chapter 11.)

Beyond the motorway with its scenic splendour of Cymric mountainous country on the right and its industrial sprawl of heartbeats on the left, the way lay along straight stretches toward Briton Ferry, where once again I had a choice, either to divert to the left, crossing the Briton Ferry bridge, or to continue on through the urban route, with chapels, hospital, tiny shops, and houses in evidence along the road all the way to Neath. If I took the urban road, my mind would go back to the days when Lin had spent time at Penrhiewtyn Hospital when we lived in Bridgend, and she required postnatal attention in 1948. The hospital is now the Neath General. If I went over the bridge, I thought of the day I had visited it when I was R.E. for the Usk pipeline in the early 1950s, when the Institution of Civil Engineers organised an official visit to see the constructionwork of the bridge. When I arrived that day in my old Rover fourteen-horsepower saloon, Bill Carlyle and John Edmunds, my young student engineers at the pipeline to gain experience to supplement their university degrees, spotted me and ran pell mell to meet me. "Mr. Price, Mr. Price . . ." They bubbled with enthusiasm to tell me of their pleasure in seeing the bridgeworks, just like two boys in school running to tell a favorite teacher about a new game just started. That had been a pregnant moment for me, a moment when my work as R.E. had meaning more than any public acclaim.

So, passing Bridgend, where my daughter, Christine, was born in November 1947, passing Abbey Works, where I had worked at the time, passing the Briton Ferry bridge, which I had seen being built in the early fifties, or passing the hospital way, I felt at home when travelling the forty or so miles from my home in Barry to the job in Neath. The journey was never boring. Traffic could be so busy that one was preoccupied simply trying to avoid it. When traffic was light, I seldom resorted to listening to the wireless, as I had done on the A47 when on the fifty-three-mile trip home from Peterborough to Leicester in 1963. The reason for this was my absorption with learning the Welsh language. I would use a hand tape recorder to talk into and play back to see if it sounded right. The strange voice speaking in Welsh was, at first, a bit hard to take, but with practice, the words seemed to come over with a bit more conviction, and it filled in the time.

On my first day on the job, Wednesday, the fourth of January 1978, the holiday break being over, I was able to meet the people with whom I would be working. After calling at the Civic Centre Building to meet those of the borough engineer's department staff concerned with my presence at Neath, I sidled down to the job, where sectional offices had been set up in a corner of the yard of the old unused pickling factory for the R.E. and staff and where the contractors, Leonard Fairclough and Company Ltd., (no relation to Len in Coronation Street) were accommodated in offices and workshops.

The compound was at hand and ready-made, though the accommodation was limited, with a pervading and distinctive smell, probably a mixture of oil, pickles, and rat muck, as the rodent officer had to make several calls to flush out the rodents. My offices comprised four rooms, a kitchen, and a conference room. They were not a pretty sight on my debut to the site, but they afforded more cover from the weather than I had started with on the Usk pipeline in 1951 (chatper 12), and they were warmer than the contractors' agent's accommodation—that was like a ruddy refrigerator, but then, contractors have to run round to get warm.

Keith Ankerson was out on the site. We shook hands, and I invited him to see me as soon as he could spare a moment. He was trying to progress a short length of deep sewer that had to be built to divert the town's main effluent sewer clear of the new works. He stood at the side of the trench, not overjoyed at what was going on. With the equipment he had it was a bitty, unglamorous process, supporting the ground with boarding and working slowly forward pipe by short bleeding pipe. The silty ground did its best to ooze into the deep trench, succeeding here and there, Over at the other side of the site near John Owens' property, the steel piling was being driven by the piling specialist Thomas Vale Ltd. of Stourport.

In my offices, a lone Sri Lankan engineer was knee deep in toil, his own private papers disclosing his itinerant life packed into a corner, contract plans lying feverishly on the drawing desk, boots, socks, and site dirt, not all top dirt, carpeting the once patterned lino floor. Sri Lenga Chellappah—it took me some time to get the name right—had no colleague on site that day save me. The clerk of Works, whose function it was to oversee the day-to-day work in the field, had not turned up. His name adorned the door next to Chellappah's: Jock Douglas, Clerk of Works.

On my fleeting visit into the offices when I came to the site on the day of my interview, I had spent ten minutes talking to the clerk of works. He was a most cheerful individual and, typical of the breed of clerks of works, methodical, interested in doing a job properly.

"I am certainly glad to meet you, Mr. Price. I do hope you'll get the job, I am sure we'll get along together," he'd said.

I went into his office. It was clean and tidy. His desk had some papers on the left hand side. There was a list of items for attention, with some of the items ticked off. A plan desk under the window had a folio of drawings of the works. Underneath, a pair of gumboots with thick white stockings and a boot jack were neatly to one side. The layout plan of the site was neatly pinned to the wall, and a filing cabinet stood by the door, with a number of files, labelled in the C.O.W.'s hand writing.

The engineer, Sri, appeared at the door. "He won't be coming in," he said with a flat expression. "Jock died over the holiday." It seemed that Bill Douglas, a man in his fifties, had had a heart attack.

There is a sense of finality in death. One cannot change the situation of the absence of the person, merely try to fill the gap. At such a time, one does not ask for a replacement. One knows that will happen in time. At the moment, it was not important. The sadness of the New Year now hung round like a cloud.

Malcolm Thomas came down from Pontypridd. Bill had been with him at the Pen-y-bont Sewage Works scheme. They had evidently been quite close. The reading of Jock Douglas's papers took me back to the time I had read about Leigh Hunt's demise when I was in Singapore. (See chapter 9.)

Very soon after I started at Neath, Ken Taylor introduced me to a secretary the borough had taken on to work with me at Quay Road. It was Mrs. Mary Rees. Sri was also in the Civic Centre at the time, so I asked him to take the lady down to the site and show her the office while I collected some data at the Civic Centre. When I returned to the office on site, Mrs. Rees was sitting in the fourth room, which she had tidied up for herself. We spent some time sorting out the arrangements for the office filing and procedure; then, with a decent cup of tea, the offices began to come to life.

In the nature of things, we were never overworked officewise. It was a job to work things out, get them done, not to conduct a great correspondence course, yet Mrs. Rees maintained a continual good humour whether she was busy or slack. She too was pleased to return to work after two years as an architect's widow. Her job was for her what mine was for me, although she was worse off without company at home, and had clearly lost weight during a prolonged period of mourning. Watching this lady enjoy her work experience and gain a little weight and overcome some of her sadness was, for me, just as rewarding as seeing the solution of the works problems that arose on the site.

It would have been impossible to work so closely with so few staff necessary for a small works of that kind without having some awareness of their personal worries. The protectiveness of the men on the site for Mary was wonderful to behold, and I am sure the atmosphere was as good as a convalescent home for her.

Keith Ankerson, agent, was really a tunnel engineer and was destined to be transferred to such a site, but he remained at Neath for the first formative months, during which period the difficulties of obstructions in the ground became apparent. At the very outset, it was impossible to get some of the cofferdam curtain piles right down to their correct depths. The scheme idea was to vibrate them into place, to avoid the shock of percussion driving to the adjacent buildings. Unfortunately, the vibrating technique was no match for the obstructions in the ground, so Director Vernon Willey could only suggest on behalf of the contractors that they use a heavy piling hammer for the difficult piles. This was to be the form for several weeks, with the vibrating equipment supplemented by the heavy hammer as necessary. Even so, it was still not possible to penetrate the full depths in some spots, so that pits had to be sunk to root out the offending objects. These turned out to be old trunks of trees, which had been on the riverbank decades ago, and also the remains of an old railway sleepered track, signs of which were ever present in the excavations for the cofferdam as each cell was dug out by skip or dredging, this excavation operation proceeding from the month of March 1978 onwards.

Meantime the piles for the support of the motor room were driven. Over that area of the site the ground was found to be too weak for the specified lengths of piles to support the loading, so that greater lengths of piling was necessary, into a deeper, firmer stratum. Every pile was closely observed during driving, and tests were made to guarantee the load bearing.

June came and the divers, a team of six men, commenced work on cleaning out the bottom of the subaqueous excavation, followed by their guiding the end of the pipe from the pumped concrete, under water to plug the bottom below the level of the reinforced concrete structure to accommodate the Archimedian screw pumps. By the end of the first year, the floors and walls of the great sump were approaching ground level.

March 1979 came around, and test loading was carried out on the sump. About two thousand tons of ballast, sand, and steel ingots from the Llanelly steelworks were run in and the test load allowed to stand to prove the support value of the ground on that half of the site. Eventually, with a very healthy performance under load, the ballast was all removed and construction of the whole pumping station motor room and generator room and ancillaries proceeded to near completion by the end of the second year.

The completion of the work at Neath was a tribute to hard working contractors' men on site, led for the most part by Nick Daines, who succeeded Keith Ankerson when Keith was transferred to tunnelling work at Manchester at an early stage in the pumping station construction.

On the tenth of April 1978, we on the official side were fortunate to have Clerk of Works Jim Clegg join us from West Glamorgan C.C., with whom he had been working on the M4 west of the Port Talbot Bypass. Jim was a tower of strength, maintaining a sense of discipline in fair weather and foul. Especially where the concrete was concerned, his compulsive delight in his work and his conscientious persistency were evidenced in the final results achieved.

Sri, the assistant R.E., kept up his end, checking out the reinforcement, keeping records, working out modifications, and keeping au fait with the setting out of the works. Junior engineers were given the opportunity by Neath Borough of gaining site experience each for six months while they studied with the Treforest College of Technology, and their efforts and enthusiasm were commendable.

For my part, although I found it very enjoyable having that two years' lease of working life, it could not alter the fact that "all work and no play makes Jack a dull boy," and some complementary activities had to be pursued.

Bob Davies' first-year Welsh course concluded in June 1978 with a convivial night out for all at the Sportsman's Rest in Peterston Super Ely and for some at the medieval banquet at Cardiff Castle, with dimmed lights, cawl in wooden bowls, mead in wood chalices, and a heavy meal on wood plates served by wenches flouncing around the tables at which sat the guests on forms, entertained with sweet ballads by the self-same serving wenches.

In the summertime, Porthkerry Park is a nice place to while away an hour. You can walk to the pebbled beach and amble around, or you can walk up the lane under the railway viaduct and emerge on the main road to Rhoose. You can stroll through the woods along a nature trail restored with loving care by the warden, Mr. Nigel Smallbone. You can also go round an eighteen-hole chip-and-putt course, nicely tended by the warden's staff, and if you feel like driving a ball a bit farther, you can choose your time to practise driving on the long parkland nestling in the woods. If your golf ball is lost in the brook, you can possibly find someone else's lost the day before.

Barry Island hardly needs a commendation from the likes of me. When our grandchildren came down, they needed no urging to visit the beach or waste their pocket money in the fairground, with all the challenges of that world. Bumping motors can be great fun with Grandpa's arm holding you in tight.

In Romilly Park there are swings for children to play as the adults walk round with the dog or without.

Penarth and Sully are not far away, and Cardiff is as nice a city as any to blow a few quid on things you did not come with the intention of buying. How exciting it is for children to board a stopper train at Barry with their parents and grandparents and toddle up to town, with people jumping in and out of the carriages, which have doors for every bench seat, warm and snug as it sizzles along behind the diesel engine.

These were the family week-end pursuits in the summers of 1978 and 1979, and frankly, what grandad could ask for more in a quasi-retirement situation? Was it the onset of second childhood? Was it preparation for more philosophical acceptance of life after 1980? Whatever it was, they were simple, enjoyable times.

What about my Welsh learning? In September 1978, I joined evening classes at Glan Ely School, the principal of which was Ieuan Jones. Goronwy Jones was the teacher, and he had a technique as different from Bob Davies's as chalk was from cheese. Goronwy's fluency in Welsh was of a kind similar to Norman Pereira's in Spanish, and he spat out his lessons at the same rate of knots. Some dropped out after a few weeks, probably disheartened. About a dozen slogged away. I was fortunate to have my periods of travelling to and from work talking to myself and was determined to stick with the O level course. During that winter, a social evening function in the Paget Room, Penarth organised for Welsh learners, with hardworking Gwylym Roberts as the emcee, resulted in Lin (who speaks Welsh) and I being invited by a young lady Cymrodorion member to join the group in Barry, which we did, and later I was invited to join the men's dining club Cylch Cinio Cymraeg,y Barri a'r Fro. These evenings provided refreshment of mind and an entirely new atmosphere for me, with personalities having a new sense of humour and values.

On Easter Saturday, April 14, 1979, our black Labrador, Cindy, ten and a half years old, died. She had been a great comfort to Lin during the times we had had to be apart for work reasons, and Lin took the loss very hard. The poor dog had not eaten for a month and had a tumour in the stomach, so the vet had to put her down. That day we could not face sleeping at home and decided to travel to North Wales in the car and pay a call on my relatives. We went up the east side of the principality, through Hereford and Shrewsbury. We stayed for two nights with my mother, who still grieved over losing Dad in June 1976. We saw Chester Cathedral and then we returned via the west side, having lunch at Bala, with its beautiful lake, seeing the national museum and the University of Aberystwyth on the way home.

Week-end visits to St. Fagans (the folk museum of Wales) and a boat trip across and up the Bristol Channel as far as Portishead, Bristol, and the river Avon helped to take Lin's mind off the absence of Cindy in the remaining April weekends that year.

My O level Welsh oral test came up in the Barry College of Further Education on May 4, and a special Welsh class in Cyncoed claimed my attention on Saturday, May 19, with Gwylym Roberts always an active worker at these functions. I think Gwylym deserved an honour for this work in promoting the Welsh language, even if so many monoglots consider Welsh a dead language.

On Thursday, May 24, Lin and I joined a coach arranged by the principal of the Glan Ely night school, Ieuan Jones, to go to the Chelsea Flower Show. It was a very good idea and an enjoyable outing, but when we were at the flower show it rained cats and dogs. So we came away not having seen Chelsea at its best and with a rather jaundiced view of the wonderful displays.

With term over, eight of us in Goronwy's class took our O levels and eight of us passed—a red-letter day for Ieuan and Goronwy.

July was marked by a surprise visit from my sister and brother-in-law Iris and Ron, whose home is in Llandudno. They had been on a caravan holiday in the West Country. We were pleased to give them a quick "cook's tour" of the area round about us.

Later that year, Lin and I took a week's holiday touring the south coasts of England and Wales and then decided that, with Cindy gone, we would change the Renault 6 car for a saloon, registering the new Renault 18GTL on August 1, 1979. Later, August was marked by the return of our daughter from the Carribean island of Dominica. She and her family were fortunate to get away just before the arrival of the devastating hurricane David, which was immediately followed by Hurricane Frederick, strangely, the two Christian names of our son-in-law.

In September 1979, I learned of a university extra-mural Welsh course, and Chris Rees, in charge, accepted me to join his class on October 2. This study was going well. It was a class where no one was allowed to speak English. It was a course for the mastery of the language, *Cwrs maestroli*, but one night it stopped.

On the eighth of January 1980, I was returning from my class in Cathays and at a quarter to nine I was driving home up the Cowbridge Road when I was involved in an accident involving three cars. My car was off the road for a month, leaving me without means of getting to night class and having to travel to Neath by bus and train until the end of January, when I was due to be released from my work there.

It had been a two-year spell of quite stimulating experiences, and I am sure I was a good deal wiser for it. The work at Neath, the evening classes, and the contact with Welsh associates and a mixture of leisure activities and pastimes had all been good for me.

From now on I would regard myself as fully retired and, do you know? The three years since then have really flown by. I'll end my story with just a few more pages to bring us right up to date at the time of writing about my progression through life.

The grass has not been growing under my feet during the last three years, in which I have finally come to terms with being "out to grass." For one thing, I am thankful to have made good use of my camera, given to me by my brother-in-law, Howard before he passed away twelve years ago, in April 1971. This provides an interesting record. A habit of sticking dates on papers and some three hundred snaps, with many cards and brochures of our most recent days, enable me to sail in with some ease to the end of my sojourn through life to April, 1983, which coincides with typing out the four hundred or so manuscript pages of this book.

Physical work, family visits, holidays by car, train, and boat, and other outdoor and indoor pursuits have filled the time of Eluned and me. We could not have asked for a better place to settle than Barry, with all the kaleidoscope of life to be savoured from here.

Lin is a Southwalian, I a Northwalian, rivalry enough to keep any marriage alive while there is breath in our bodies. And while Lin forever yearns for a life in which her family live close at hand, as do some families, I have long had to come to terms with the disparate lives of my own side of the family and of our children, who have each spent so much time abroad. But when you live together as long as Lin and I have—we are now in our fifth decade of married life—the fervour of the one's pursuits inevitably rubs off on the other.

Also, I have always enjoyed driving a car. Lin hasn't got a clue where road sense is concerned, though more often than not she gives unsought advice in the car on taking shortcuts up the wrong way of one-way streets or making a turning into a side road we have just passed. In my mature years, I take great delight in chauffering her around to the women's circles, shopping, and down to the beach for a walk in the sea breeze at any one of a dozen spots on the South Wales coastline, usually not too far away. At the same time we eat regularly and tastily at home, cooking at regular hours, with me doing a fairer share of the washing up and siding away the tea things and always taking a cup of tea or coffee up to bed before we get up in the morning, which all makes for a harmonious routine agreeable to body and soul.

It takes a very long time to get used to the idea that one's time is one's own when one has lived one's life as a product of society implanted with the work ethic. Any time I had taken off during my past life in engineering had been accompanied with a sense of guilt, of playing truant. Even to this day my conscience is really being salved by the fact that, having turned to writing, I may be doing something that someone somewhere will regard as having some value. No more can I leave behind new constructionworks to bear witness to my current efforts, in the way I seemed to regard those jobs for which I made my contribution in the past. I can only go on working on paper. Maybe I will thus do something worthwhile.

Early in February, 1980, I had immediately burst free of engineering. Lin and I went off to stay a week in Guildford with Christine and her family, my sixty-third birthday on February 6th making an excuse for a get-together. We could also take care of our two young grandaughters while their parents went off for a break in the West Country. On Monday, the eleventh, we also managed to see our son Newton, who was working in a bank in London. A day out in London Town was something of a delight for us, as it had been some time since we had felt free to visit the metropolis. As it was, we enjoyed our son's prolonged lunch break looking round well-known landmarks—the Stockmarket, the Royal Exchange Buildings, and Saint Paul's Cathedral.

By the twelfth of the month, we were ready to return to spend the rest of February experiencing the novelty of so much free time. I spent several individual days digging and planting the garden, and on Sunday, the seventeenth, we decided to enjoy a lovely day outdoors, visiting Dunraven Castle near Southerndown, on the lovely South Wales Heritage Coast, one of the first lengths of the coastlines of this country to be claimed as such, reserved for coastal walks.

February went out on a high note. At Mount Sorrel Hotel, Barry, Lin and I attended the annual dinner of the Welsh Society, the "Cymrodorion," to celebrate St. David's Day, March 1.

March came in like a lamb and Sunday, the second, found us preparing for a tour of the West Country. The next day, we set off from Barry at ten past nine, travelling under a dull sky along the Welsh side of the Bristol Channel, crossing the Severn Bridge, and calling in at the Aust service area for a cup of tea. We went on via the M4/M5 junction travelling south on the M5 past Bristol to Taunton, ninety miles from home. We stopped at Taunton for lunch, taking time to look in at the Church of Saint Mary magdalene. Taunton is twinned with Lisieux. Leaving Taunton behind, we took the A361 westward, enjoying the long country stretches and the small villages. Passing through Barnstable, we continued on our way along the A39, to cross the bridge over the river Torridge at Bideford, where we stayed the night at a guest house near the waterfront.

The next morning it was sunny and cool, and making a sortie to buy a paper, I saw someone remarkably like Angela Rippon driving a small car in the streets of Bideford. Breakfast over, we left Bideford, taking the road to Northam and Westward Ho! Everything was closed for the winter when we looked in on the beach, fresh in the bright sharpness of the morning. Town shops were open, and council employees were preparing for the spring season. We left Westward Ho! behind us, rejoining the A39 to head south, with periodic views of the sea, until a sign pointed right for Tintagel. The boundary sign Kernow heralded our arrival in Celtic terrain, and I determined to see if I could acquire a dictionary to compare the Cornish language with my understanding of the Welsh language. When in the coastal town of Tintagel, we visited the six-hundred-year-old post office although, it was officially closed. This building is held in National Trust. The old building has its Parliamentary Clock and viewing gallery with the original furniture.

After leaving Tintagel, we again joined the A39, this time at Camelford, and continued our way on the main road until we crossed over the fifteenth-century stonemasonry bridge over the river Camel at Wadebridge. Lunch in the Molesworth Arms was a delight. The quaint inn had a comfortable dining room and first-class cuisine. I had a very filling and enjoyable home-made beef curry, and Lin enjoyed her roast beef, Yorkshire pudding, and vegetables. Feeling well satisfied, we motored on from Wadebridge to Newquay.

The car was parked in the public car park near the railway, and as we walked round the town, the Atlantic air was so bracing we were tempted to stay the night, but with a fresh pot of tea and cakes inside us, we journeyed on down the A3075 to meet the A30 at Blackwater, rolling on as far as Penzance.

Our second overnight stay was therefore in the Sea Front Stanley Guest House, with Mr. and Mrs. Dyer. There, if so inclined, one can leave one's car overnight in the cark park and travel over to visit the Scilly Isles on the Scillonian ferryboat or on a helicopter from the heliport at Penzance. Of the many antique buildings in the town, perhaps the four-hundred-year-old Benbow Tavern conjures up most nostalgia as the one best known, with its *Treasure Island* association.

The rain on the morning of Wednesday, the fifth of March, was washing the seafront promenade outside our lodgement as we left after breakfast, along the B3315 coastal road, through the fish market of Newlyn. Our final leg of the journey to Land's End was just 15 miles, and when we got to the end of the road we had completed 290 miles since leaving home. Land's End was very blustery and wet, and we simply dived into the hotel, buying souvenirs, then braving the elements outside to take a quick look round the rocky cliffs and taking a picture or two before turning round to head back from whence we had come, as it crossed my mind that at the other extreme of the A30 lay Ashford, Middlesex, and the Bulldog improvement, referred to in the previous chapter.

We decided to return along the rain-lashed roads to Penzance, passing through Helston circuiting around in congested Falmouth, and breathing a sigh of relief when, at last, the sun appeared in time for us to pull up in the forecourt of the Norway Inn, nestling back from the road at Devenish, where we clicked for an appetising bar lunch of steak and kidney pie and vegetables. While the sun still shone, we looked in at the cathedral at Truro. Then we walked round the shops, and it was there that I was able to find books in the Cornish language, Kernew, a flick through the pages quickly suggesting similarity of the words in Welsh, particularly the numerals. We had been hoping to make an overnight stop on our return, but the weather was against us at places like Falmouth, Plymouth, and Exeter, where we had thought of staying, with heavy rain falling in each town as we passed through.

Finally, we pulled in for a snack at the Granada Service Area at Exeter, the southern end of the M5 before resolving to return home through the black night in torrential rain, arriving back at the house at half past nine in the evening, having had enough of a run that day for me to need no rocking to sleep.

Our trip to the West Country came about because it was the first of the places on a list I had prepared at the beginning of 1980, places we felt like visiting or revisiting during our retirement years. Other places on the list were the Channel Isles, Britain's coastline, the heart of England, Spain, the Mediterranean, Italy, France, Belgium, Germany, Holland, Denmark, Sweden, and Norway—not much if you say it quickly.

With such a list of possibilities on the agenda for the future and with a backlog of work to be done around the house and in the garden and with the forthcoming visits of our children and grandchildren and return visits inevitable, who could be bored with life? After undertaking our other commitments, 1980 was to see us with time to make only three visits on the prepared list, which only goes to show how time flies. Apart from the Cornwall outing at the beginning of March, we squeezed in a three-day boat trip to Jersey at the end of April and a week's tour round the heart of England in the autumn.

Some men, and women, too, for that matter, can spend their whole life in the garden. Maybe I will find myself doing that in the future, but so far I am not entirely fulfilled in that pursuit. Maybe my genes are still ticking over in tune with the outside world, as I still seem to be reaching out for other parts of the spectrum of what life has to offer, outside my known experience. Fortunately, Lin is game to venture afield quite frequently provided it is somewhere she wants or can be persuaded to go, and so, with compromise, we both find pleasure in travel. But most of my serious time during March 1980 was spent in gardening, ordering and preparing timber for the superstructure of a greenhouse (for which I had erected concrete walls while I was still working for my living), and pottering about as Lin's chauffeur.

There came a *cri de coeur* from Christine for us to go up to Guildford to look after our grandaughters for two weeks while she pursued a hospital course in midwifery. This working holiday was in April; with Christine having but fleeting family time, Lin did most of the domestic chores in the house. During this time we did manage a spin out, one Saturday, the twelfth, from Guildford to Sevenoaks and Tunbridge Wells, to renew old memories of the area in Kent, with Knole Park, lunch in the Tudor Cottage, and tea at Binns, a Tudor-style restaurant in the wonderful Pantiles in Tunbridge Wells. Returning to Barry, it was down to the job of rebating the timbers for the glazing of the greenhouse and getting the garden straight.

It was Monday, the twenty-eighth of April, when we started off on the Jersey trip. The four days away from home seemed like a week's holiday as we savoured the car trip to Weymouth, the Sealink trip on the Earl Godwin, three nights' stay in a St. Helier guest house, a coach trip round the island, and a sea outing by hydrofoil on the Wednesday to the port of St. Malo, solidly built, with stone streets protected by high walls. There I was able to buy a book in Breton, the Celtic language of Britanny, with its associations with Cornwall and Wales. We shared a table for lunch with a couple from Paris, and the wine and the fish main course and the company were sheer satisfaction. Returning home from Weymouth by car on the Thursday afternoon, we made our break complete with tea and scones in an olde worlde restaurant in Dorchester.

In the light of my retirement, all our family were moved to see us in the fine-weather months of May, June, and July, and what with family visits to and fro and sightseeing trips with friends, we had an almost bewildering bevy of busy days. When our son David, wife Jane, and Gina and Jeff flew over from Minneapolis, they had to divide their time between Kent, Surrey, and Glamorgan, where dwelt respectively our older son, our daughter, and us. Time just flew along as we tried to make up for those nine years we had not seen our son David. Then, of course, when he and his wife Gina and family returned home to the USA from Gatwick on August 21, there was not a dry eye between us. David had spent his time in Wales reliving his schooldays in Pembrey, walking along the paths he had walked as a boy to the old Pembrey harbour and to the docks in Burry Port, where he had dived in with his pals in those far off days. He was able to take away from St. Illtyd's Church a recent publication of the church history and from Penarth and elsewhere a number of souvenirs to make up a "Welsh room" back in Bloomington, Minnesota, U.S.A.

As soon as David returned to the New World, I proceeded to make progress with the greenhouse superstructure, a strong affair made from solid timber and stout glass with Georgian wired glass panes for the roof. The months of September and October seemed to pass by unnoticed, me pottering about in a heavy sort of way and Lin always there, coaxing colour from her garden flowers while from time to time we should go for a spin for food or to look round the area. Then, with the construction of the greenhouse well and truly finished, including its painting and, for good measure, some painting of the house, we decided to take that "heart of England" tour.

This was a five-day tour in the car from Monday, September 29, until Friday, October 3, inclusive, and it was to be a very nice holiday before the onset of the backend of the year.

The route we picked as representative of the heart of England was to take us to Gloucester, Oxford, Cambridge, Bury St. Edmunds, King's Lynn, Spalding, Melton Mowbray, Nottingham, and back home through Shrewsbury, Welshpool, and Merthyr, really a heart of England and Wales tour. These are not just places on the map. They all have beating hearts, and anyone with a car can discover the wonder of our country simply by going off for a few days and staying the night here and there, without too much preplanning.

On our first day we stood by the tomb of William the Conqueror's son, Robert Curtoise, duke of Normandy, in the magnificently colourful Gloucester Cathedral. At Banbury Cross, we pulled up and thought to stay for the night, but the prices at the Moat House were not to our liking, so we made our way on the A41 and A423 in the direction of Oxford, stopping when we saw a roadside sign "Bed and Breakfast" in Deddington. The particular guest house was full, but we were given the name of a lady who offered bed and breakfast, at her farmhouse. After losing our way up a lane and again being redirected, we were invited into the glowing farmstead home of Audrey Fowler. The real country home welcome was just what we needed in the chill of the evening. An enormous woodburning stove, French in character, stood proudly sited amongst the soft furnishings of the old farm lounge to greet the welcome strangers. Tea came in, as free as the smile on Audrey's face, and when she took us upstairs to our room we were at once in the world of the heart of the country.

After a restful and quiet night, we awoke to the sounds of cows mooing and chickens being fed. The man of the house was energetic and friendly, a worker as well as his own boss. He it was who had made the new swimming pool near the house, and before we left I took a picture of Lin and Audrey together behind the pool and with the house in the background, with garden walls softened by sweet country verdure and flowers, which still bring the moment to life, jumping out of the picture.

Oxford is but a cock stride away from Deddington, about half an hour's run, and ten o'clock in the morning saw us exploring the delights of the wonderful university city. For heaven's sake, do not leave this life without going to Oxford! It is a beautiful city with a beautiful heart, and from the top of the cupola of the Sheldonian Theatre you can see the panorama of principal university buildings. Before leaving Oxford, we purchased a beautifully-illustrated book on the city. This was just right as a present for Eluned's cousin Margaret Jones, a retired schoolmistress, just recovering from an abdominal operation at the Heath Hospital.

On then to the next leg of our journey, and we were heading for Cambridge, that other ancient seat of learning. You cannot get to Cambridge without crossing that great wide way, that pioneer of motorways, the M1, so familiar to me because of the part I had played as its chief resident engineer over two decades ago. We stopped on the overbridge carrying the A418 between Husborne Crawley and Ampthill, and I took photos of the majesty of the highway curving up to me from London and heading north, straight as a die, to vanish far in the distance.

My mind returned to those days of chapter 14. I could have been meeting Arthur Price, my R.E. for contract A, or Vic Poulton, who was R.E. for contract B, as we were close to the junction of those two contract sections, but a glance at the weeds growing in the bridge tarmac footpath snapped me into the present. I knew my thoughts to be fanciful. In any case, Vic is no longer with us on this earth.

Arriving at Cambridge was a bit of an anticlimax after Oxford. Perhaps it was because I always did have difficulty finding my way through that city. Many a time when we lived in the Midlands, returning from the east coast via Haverhill, I had come to the T-junction at Trumpington and then proceeded to lose myself. Now the outside sprawl of the north-west corner of the city was upon me and, tired as we were in the late afternoon, we did not see the place at its best. We settled for a cup of tea and a sandwich in the railway station buffet and then struggled on again along the A45, through Newmarket, to find ourselves a place for the night at Bury St. Edmunds.

The guest house was just opened, looking spic and span. The inside of this large terraced house was modernised. It was not particularly warm as, in the chill night air, the heating was hardly adequate, but that was not so important, as we were in a mood to move around a bit to overcome our inactivity in the car. There is a power in the atmosphere of the borough of St. Edmundsbury, with its motto "Shrine of a king, cradle of the law." The official guidebook is a joy to behold. If you want to learn something about British heritage and freedom, then go to St. Edmunds, see the cathedral and the abbey, and be amazed at this treasure of our land. I will say no more about it, for me it is indescribable.

So, on the morning of October 1, after our absorption with the ancient settlement, we took the A134 to Thetford and, driving in the woods for a wayside picnic lunch, continued on, stopping to walk round

King's Lynn before moving on to our destination for the night, Melton Mowbray. Melton Mowbray had its large and beautiful parish church of Saint Mary the Virgin, the Corn Exchange, Melton Mowbray pork pies, its hunting, and a house claiming its one time ownership by the young Anne of Cleves, one of the wives of the much married Henry VIII, all to remember it by. We booked in at the Westbourne.

I wanted to see Belvoir Castle and Nottingham, so the following morning after a quick look around town and then waiting awhile after breakfast until the ground frost had disappeared, Lin and I followed the A506 to Nottingham. Belvoir castle is on the way, but it was shut as we went by the turning that Thursday morning, October 2, and so my next objective was to try to find the lecture room where I had addressed a body of students, lecturers, and a variety of engineers, some 220 strong, on November 30, 1960, the subject being the M1 construction which hovered in the immediate future for so many of the audience that night.

I had a longstanding insurance policy with Sun Life, but when we got to Nottingham, I found they were no longer in offices at 9 Low Pavement and could not trace them nearby. The Nottingham Theatre Royal claimed our attention for a while. Then we walked round and round the city, seeing what it had to offer, losing ourselves over and over.

We simply had to go to the polytechnic to find that lecture hall. Vast extensions and alterations had been carried out since I made my visit in 1960, and it took some time to locate the "small" lecture hall on the first floor near the Newton Building entrance of what is now known as the Trent Polytechnic. None of the lecturers I met were able to help me with my inquiry, as no one seemed to go back so far, but an astute student who seemed to have particular interest in the poly's origins appeared and so I stood at last on the spot I had delivered my talk, while Lin took a photo for the record.

One more night spent at the Westbourne Hotel in Melton Mowbray and we took our leave of Marion and Jim, our hosts, to make tracks in the direction of Leicester, where we suddenly realised we had our room key with us and posted it back to the Westbourne. Once we had made it through Leicester, we had the bit between our teeth and decided to make the journey home in one day. Lunch was at Ashby-de-la-Zouche and tea at Shrewsbury, with its Welsh and English bridges, and the car seemed to be taking over as we sped along on well-known roads through Welshpool, Newtown, and Rhayader, where fish and chips in the car in the gathering dusk made a welcome break. Builth Wells, Brecon, Merthyr, and back home by half past nine was the end of a happy but exhausting round tour of seven hundred miles.

When you return from a week's holiday like that, you need another week to get over it. However, something had to be done about the front driveway to our humble estate abode. People visiting had been trudging across the lawn at an angle to make a beeline for the front door, wetting their shoes and spoiling the green sward of squelchy grass, and it offended my sense of order. So on Saturday, the eleventh of October, I set to work making a turn in, a sort of lay-by, to give some more paved area in front of the house. It was heavy excavation, wet, sticky ground, and I took off a goodly layer of the stuff and wheeled and tipped it tidily at the end of the garden behind the house. Then came the cement bags and the aggregates and sand and all the hand mixing, made worse by the fact that the hole filled with water, our property at the front being without fall in any direction. Using my body as a concrete mixer after bailing out the water, I had a few hard days getting in the base, finally incorporating a neat surface of concrete tiles in the top layer of concrete. I was pleased with the result—no pitting, no failing, no hanging water.

The year 1980 closed on a strange note for us as a family. From Christine's birthday, the thirteenth of November, to a short while before Christmas we had been keeping our daughter company, as David was then working abroad in Nigeria. We learned that the young family would be united overseas over the holiday, and so we had agreed to have a mock Christmas meal together beforehand. As we had an invite from our old friends Harry and Mary to visit them at Christmas, all boded well. When we had had our premature celebration lunch things went awry in Africa, with Christine receiving a message not to go after all. Then, after the tickets had been cancelled, once again came a message that it would be all right to make the trip, but that was too late to arrange.

When back at Barry, we were about to start out to our friends in the Midlands, the weather was too foggy, so all plans for the holiday were upturned. So we all finished the old year on a strange note. "That's the way the cookie crumbles" seems to be as good a proverb as any to fit the situation.

The new year of 1981 began more cheerfully when our daughter and grandchildren were down for a week's holiday. Since then 1981 and 1982 seem to have gone in a trice, yet quite a lot has happened while we settled down, more and more, into retirement. Humans seem to grow into a place like conifers into a hedge, and since we planted most of ours a foot high in 1980 they now average about five feet high. When we look at them, we seem to belong where we are.

Summarising the highlights of the last two years; in 1981 we had ten days in Guernsey in the April; a visit to the Strand Theatre in London, on the twenty-fifth of June, to see *No Sex: Please, We're British* that longest-running comedy, written by Anthony Marriott and my old colleague Alistair Foot; and during the last four months of that year, the erection of a small rear porch to act as a storm protection. In 1982, March 24 saw us hovercrafting over from Dover to Boulogne in the *Princess Anne*, enjoying the old walled cathedral town of Haut Ville and a "Civils" visit to the site of the realignment and reconstruction of a particularly tough piece of the A467, from Crumlin to Aberbeeg, these trips offering refreshment to the mind, each in its own way. In May we made a visit to Pembroke, and on the second of June Pope John Paul II passed over our house as he flew from Rhoose Airport to Cardiff, returning by Popemobile on the B4265, past the end of the road just a hundred yards away from the house as he made his way to leave this country from the airport. Later in June, a few days in Jersey saw us photographing some of the Bergerac cast on location, and enjoying a chat with that old English gentleman, and comic Tommy Trinder, hard-working as ever at seventy-three. August and September saw us enjoying special week-end holiday breaks by British Rail, and our last leisure car trip was to Salisbury, Old Sarum, and Broadlands at Romsey, the former home of Lord Louis and Lady Edwina Mountbatten.

From the foregoing pages of this episode, during the last three years you may agree that the grass has not been allowed to grow under my feet, even if my accomplishments are nothing to crow about, time having been a fleeting friend for me. One is always learning, and I have appreciated the flavour of life's new experience as a member of Barry groups like the Club Cinio, where Islwyn Jones is an ever present source of jollification, and the Forum whose founder-member and late president, Willie Carpenter, was buried only yesterday, the seventh of April 1983, a gentle soul of over ninety years, with Chairman Len Gerry, person extraordinary, being the meetings' guiding light for the last eight years, his time given generously to the organising involved in getting a series of first-class speakers on every subject imaginable, politics and religion excepted.

Each year Glamorgan County Council Education Department publish a programme of night school courses. In 1982, Lin picked one out for me—"Short Story Writing," a short, six-weeks course, with Pamela Cockrill taking the class at Bryn Halfren School on the Port Road East. (Shades of Bob Davies.) I almost missed joining the class when I turned up on the twenty-second of September, a week after the signing on date, but fortunately, I was registered as the last and twentieth member, although a smaller number had been envisaged for the size of such a class.

During the first six weeks of the course I wrote or spoke into a tape recorder about twenty individual occurrences during my lifetime, and then, on Sunday, October 24, when we knew the course was going to be extended, I began to outline my thoughts for using these stories as milestones to mark my progression through life, remembering some of my experiences along the way. I finished the story in longhand on December 31, of 1982, and have since had to buy a typewriter and type it all out, starting at the rate of about four words a minute, speeding up to about twelve. As a writer, I am not even an amateur. I am just a beginner, but whereas I may learn to write a better story in the future, I can never relive my life. It is nice to have put something down about it at last, my own autobiography: *Price's Progress.*

James Price
C. Eng., M. Eng. (Liverpool), F.I.C.E., M.I.W.E.M.,
Chartered Engineer—Chartered Civil Engineer
"In Retirement"

Portrait by STP Photography, Cardiff.

Epilogue

The Epilogue, a Dedication to All Those People on the First of February 1977

Will I ever forget their names
Who called me Jimmy or Jim or James,
Dubbed me Jamy or Jas or Price?
How I remember them, all so nice.

In Flint, the Wood, and Ewloe Green,
Hawarden, the Varsity, Pembrey Scene,
Trecwn, and Liverpool over again,
Crossing the Mersey, by boat and train.

Will I always remember *them*, from
Ships that took me, brought me home,
The far off lands and strange folk lore,
Meeting the peoples on Earth galore?

The M/V *Accra*, Elder Dempster, with "Sos,"
Meeting young people calling us "boss,"
Harrison's Line, and in a convoy,
Safely escorted out east via the *Bay*.

Out to Colombo, up to Bombay,
Rumbling up, five nights 'n' four days,
Wondering why I had volunteered,
Leaving again when the problems cleared.

Never forgetting the navy ones, too,
LSTs, troopships, and all of their crew.
(Later, in peacetime, not flying back,
On the *Britannic*, with Woollatt and Mac.)

Hot Singapore, the snow in Rosyth,
Dusty Margam, formidable site,
Water at Swansea, Cwmdonkin, and Clyne,
Clase, Llandeilo, the Ammanford line.

Railways in Africa, M1, Northants,
Hertfordshire, Buckingham, Leicester, and Notts,
Welton and Newport and then Forest East,
Whittlesey, Peterborough, Burton not least.

What of the people down Tilbury way
Building the silos all night and all day?
What of the life in Essex and Kent,
Always delightful wherever one went?

Royal Kingston Borough, old Camberley,
Ashford and Guildford and Esher BP,
Ewell and Dorking in Surrey, two
More of the places there that I knew.

Then retiring to Barry in Wales,
The peaceful summers, the winter gales,
Eluned beside me, together at last,
Our children abroad, part of the past.

Yes, I suppose I will forget
Some of the sea of faces I've met,
But I'll not forget the worth,
Of the people I've known on earth.

Family photo of visit to Enloe Green of author's Uncle Harry after forty-five years in the United States, 1951.